JUAN DE LA CIERVA
AND HIS
AUTOGIROS

Juan de la Cierva poses with C.19 Mk.II, G-AAKY, at Bethayres Field, Philadelphia.

Stenlake Publishing Ltd

© 2011 Arthur W. J. G. Ord-Hume
First Published in the United Kingdom, 2011
Stenlake Publishing Limited
54-58 Mill Square, Catrine, KA5 6RD
01290 551122
www.stenlake.co.uk

ISBN 9781840335590

Test pilot Reginald Brie demonstrates the RAF's Cierva C.40 prototype at Hanworth in January 1938.

ACKNOWLEDGEMENTS:

My thanks to those of my friends who have provided access to material and photographs including the late A J Jackson, the late Henry Marsh, the late Alan Brie, the late Peter Brooks, the late Stephen Pitcairn. Happily still this side of the Great Divide are Bruce H Charnov, Donald A Eckel, George Cull, Richard T Riding and Mike J Hooks.

FURTHER READING

Brooks, Peter Wright: *Cierva Autogiros: The Development of Rotary-Winged Flight*. Smithsonian Institution Press, Washington, USA, 1988

Charnov, Bruce H: *From Autogiro to Gyroplane*. Praeger, Westport, USA, 2003

Ord-Hume, Arthur W J G: *The Autogiro: Rotary Wings Before the Helicopter*. Mushroom Model Publications, Petersfield, Hampshire, 2009

Frank Kingston: *Legacy of Wings: The Harold F Pitcairn Story*. T-D Associates, Pennsylvania, 1981

Townson, George: *Autogiro: The Story of the Windmill Plane*. Aero Publishers, Fallbrook, USA, 1985

INTRODUCTION

Before we start I should explain that the real title for this book ought to be something snappy like 'Juan de la Cierva, his invention of the Autogiro and what happened when it got to Farnborough and how Harold Pitcairn and the Kelletts completed Cierva's work after he was dead'. Or similar. The reason is that there are several stars in this story besides Cierva himself. Harold Pitcairn and Wallace Kellett are as indispensable to my story as are Nicholas Comper and Cyril Pullin or James Weir. That, unfortunately, is the way with the autogyro and certainly in recent years there have been important figures who never knew Cierva at all – some were not even born before he died. And they're all part of my story. I put in this disclaimer now in case you want to be awkward and point out the obvious after you've read the thing. OK?

Today's attitude towards the autogyro is generally one of benevolent condescension. Yes, sure, it was an odd idea that people followed many years ago but it was quickly replaced by the helicopter which could do things no autogyro could ever do.

Well, that's one way of looking at it, I suppose. There is, though, the fundamental truth that had it not been for the engineering and basic aerodynamic research that went in to perfecting the autogyro we would not have had the helicopter. At least, it would have taken much longer to come into being. The autogyro paved the way and smoothed its road to success. Which is not to say that the idea of the powered rotor or helicopter came along post-autogyro for proposals for making an aerial variant of the Archimedian Screw go back to Leonardo da Vinci and, in more recent times, to Vittorio Isacco and Louis Brennan. All pre-dated Cierva yet none really worked until Cierva pioneered work to find out why and how the rotor blade actually worked.

The more we look into the background of the autogyro, though, the more we find some uncomfortable anomalies in this 'old technology' business. Some of the 'old' stuff doesn't quite add up. It's a bit like comparing a horse with a car and asserting that cars replaced horses. The fact is that in many cases that is a true statement except that horses run on hay which you can grow while cars run on increasingly expensive oil products. There's also the old chestnut about the farmer who, having been told that his new car could do absolutely anything a horse could do, promptly drove it through a hedge, explaining afterwards that he'd shouted 'Jump!' at the top of his voice, but the car ignored him.

What is really worrying about the autogyro being past history is that its replacement (the helicopter) does not offer a 'like-for-like' exchange. Sure, it goes up and down vertically like a lift, hovers like a bird and even flies backwards – all things that no self-respecting autogyro would ever even attempt. But look at the cost! Autogyros were quiet(ish) flying machines making no more noise than a light plane from the local flying club. The helicopter announces its impending arrival from great distance with a sound that can vary between that produced by an incoming 1944 Nazi V1 Flying Bomb to the ear-shattering pounding of an unsilenced 1960s pneumatic road-drill.

'Dig that crazy carpet-beater, man!', cries the diligent observer looking up from the inconsequentialities of his 'red-top' newspaper as the helicopter, its rotor but a blurr as it whizzes round in ferocious pursuit of something one may never understand, thunders overhead with the delicacy and finesse of something very nasty and probably lethal in a farmyard. And when the thing hovers over you, it forces kerosene fumes (the 'in' term for these is 'particulates' – sounds much more technical!) into your hair and thumps the living daylights out of your ear-drums.

By comparison, the rotor blades on the autogiro turn slowly and more or less silently (OK: there's a mild swishing sound but who ever was deafened by the noise of butterflies in an adjacent field), while the helicopter power-drives its blades at enormous speeds to the accompaniment of a cacophony of gargantuan proportions.

For the engineer, power-driving a rotor demands some serious engineering in the way of gear-boxes and drive-shafts, pressure-lubricated bearings and an overall fear of imminent problems through metal fatigue, the risk of something going wrong, and the potential for life-long deafness for all associated with the flying and maintenance of the thing, not to mention the damage to the hearing of the untold innocent and helpless masses over which the helicopter flies.

But let's get one thing straight and it's the old one about 'horses for courses'. I do not deny that there are things a helicopter can do that no other flying machine can even attempt and were I trapped on a sinking boat in a storm, lying in a mountain crevasse with a broken leg, smashed up in a road accident, or about to give birth to my first child in a snowdrift-isolated farmhouse in the middle of nowhere, the helicopter's friendly roar would be my manna from Heaven.

Nevertheless, in terms of fun flying, crowd surveillance and point-to-point travel by air from sports fields, the autogyro is a far more people-friendly vehicle.

So why aren't we surrounded by autogyros? We know that they used to exist and that they were heavily promoted. What went wrong?

Three years ago when I produced my definitive study of this type of aircraft (*Autogiro: Rotary Wings Before the Helicopter*: see recommended further reading on page two), somebody told me it was all very well writing a big and expensive book but it probably wouldn't get seen by the ordinary person who (like me) mightn't be keen to lash out on a big book yet still wants to know a bit about autogyros. What was needed, I was told, was an affordable but intelligent guide to these curious aeroplanes.

So this is how this intentionally small book has come to be written. I offer it as a brief yet practical history of the autogyro, its invention and development. It's full of pictures (a large number of which have not been published before) and in a brief text I shall also attempt to explain the story of autogyros and where it all went wrong. In relating this story, I have relied heavily on illustrations which help to denote the progress of the machine and those associated in its development. Be warned, though, that some of these pictures are very old, many are of poor quality, and some are from my own half-a-crown Bakelite Box Brownie camera operated with a schoolboy's unrefined eye for composition (let alone detail). Pictures are chosen for their rarity, for the story they tell rather than for being potential prize-winners at the local camera club. And the captions tell the detailed story behind each photograph.

Before starting, there's an inconsistency I have to explain. An aircraft with auto-rotative wings is normally called an 'autogyro' spelled with a 'y' and a small 'a'. However, Juan de la Cierva patented the name of his invention as an 'Autogiro' spelled with a capital letter 'A' and an 'i'. Autogiro is therefore a registered trademark like Hoover and Pianola and should be respected as such. However, over the years all three names have tended to become generic terms for the devices they represent. Strictly speaking, when one talks of a machine created to the Cierva patents, one should always call it an Autogiro but all other makes of similar machine are 'autogyros'. Accordingly, I try to follow this rule in my text. Nothing is ever quite straightforward!

And another thing! What's the difference between the Autogiro and the helicopter? Simply that one has a free-turning rotor, the other a powered one. The Autogiro flies by inclining its rotor disc back from the direction of flight and, as the aircraft is pulled through the sky by the engine and propeller (just like in an ordinary aeroplane), the lift is generated by the rotor blades which are kept turning by the passage of the air passing through them from underneath to above. The rotor, like the wing of an ordinary aeroplane, is kept at an 'angle of attack' to the airflow.

The helicopter applies engine power to its rotor blades at all times so that they draw the air downwards from above. This generates lift and the aircraft literally screws itself upwards into the air. Tipping the rotor disc forwards creates an additional proportion of forward thrust so the machine travels forward. If the engine fails, though, disengaging the rotor so that it becomes free-turning like that of the Autogiro allows the aircraft to descend slowly under control as the air now passes upwards through the rotor: this is called 'auto-rotation' and, while a vital attribute for the helicopter's safety in an emergency, this is the fundamental reason how the Autogiro works.

Occasionally aircraft are designed with powered rotors like the helicopter, but the rotor can be disengaged once in the air and forward flight be sustained on fixed wings, an example being the Gyrodyne. These are called 'compound' machines for they unite autorotation (the Autogiro) with the driven rotor (the helicopter).

Some later autogyros (and this includes most modern ones) have the ability to spin up their rotors before take-off using a power off-take from the engine. This should not be confused with the helicopter because this power drive is always disengaged before flight.

HOW IT ALL BEGAN

Most great inventions can be traced back either to the ancient Arab or Chinese civilisations and the autogyro is no exception although Mother Nature was the actual inspiration. Sycamore trees reproduce by equipping their heavy seeds with a single slender stiff leaf like a rotor. In perfect balance the falling seeds can fly considerable distances from the mother tree through natural autorotation. By the fourth century BC, Chinese children were playing with a curious toy that was part shuttlecock and part sycamore seed. It was shaped like a stick with an array of feathers at one end each of which was set at a slight angle to the axis of the stick. When this stick was rotated fast using a whip or a bow, it would fly up into the air, its height depending on the angle of the feathers, the weight of the stick and the speed of rotation. And then it would slowly fall back to earth still rotating in the opposite direction.

This ancient toy was to be found in various forms for well over two thousand years but the first recorded illustration is to be found in a Flemish manuscript dating from about 1325. Other pictorial references suggest that Leonardo da Vinci, said to be the father of the helicopter, would have played with such a toy early in the 15th century. A version of this toy, made in plastic, can still be bought to this day.

Once heavier-than-air flight had been mastered in 1853 by Yorkshireman Sir George Cayley, and powered flight fifty years later by the American brothers Wright, many attempts to establish flight by rotating wings were made. Numerous contraptions were made, often verging on the lunatic: some are illustrated in my book Autogiro referred to earlier, but it wasn't until flat-disc gramophone inventor Emile Berliner applied his not inconsiderable mental powers to the task that the world's first successful helicopter emerged. On 17th June 1922 at College Park, Maryland, the Berliner flew, moving forward by actually tilting its twin rotors. Berliner and his brother Henry then, for some reason, considered they had proved a point – and abandoned the whole thing!

On 30th January 1924, Paris saw the vast sixteen-rotor machine made by the Spanish Marquis Raoul Pateras Pescara rise into the air. There wasn't much else it could do, though. The world waited until 14th April 1924 when a young motor engineer from the Peugeot car factory named Étienne Oemichen got his twelve-rotored monster to fly. He astounded everybody, himself included, by flying it in a straight line for 360 metres. The honour was short-lived because Pescara quickly came back with another and better machine.

All these were helicopters with powered rotors. There was precious little else they could perform except to get off the ground and move forward. Their success was limited because nobody really understood how rotors worked. And that is where the basic early design work on the autogyro played such a vital part.

THE SPANISH CONNECTION…

It was in 1910 that the pioneering and flamboyant French aviator Julien Mamet (1877-1932) toured Spain flying his Blériot XI monoplane to the acclaim of an admiring public for whom flight by humans was a wholly new, awe-inspiring happening. On 23rd March he demonstrated in front of the Spanish King and among those who watched Mamet that day was a young boy named Don Juan de la Cierva Cordorníu who had been born in the Murcia region of Spain in 1895. He had been around 14 years old when Louis Blériot flew from France across the English Channel to Britain and these two events convinced the young man that his future lay in the science of aviation. By the age of 16 he had built his first aircraft, a small monoplane based on the Nieuport and flown at Cuatro Vientos Aerodrome near Madrid. It was the first Spanish-built aircraft to fly successfully.

A second machine followed as meanwhile the young man followed a good college education. In September 1918 the director of Spanish Air Services announced a government-funded aircraft competition to design a multi-engined bomber. The successful design would win a purse of 60,000 pesetas which at that time was worth £2,378.5/- or at 2011 values, £63,620. Juan de la Cierva's three-engined biplane heavy bomber was the only entrant in its class and actually flew quite well but the military pilot directed to fly it (test pilots hadn't yet been invented) managed to stall it in a turn, destroying Cierva's work in the ensuing crash. Since it had actually flown as stipulated, the cash prize was duly awarded but actually covered less than half the young Spaniard's outlay. It was nevertheless a worthwhile price he had to pay for it established his name as that of an aircraft designer – Spain's first.

It was the stall of his aircraft that inspired Cierva to seek a safer slow-speed flying machine and he began experimenting with a windmill toy of the ancient type used all those years ago in China. His experiments gradually translated into his first full-size machine which he thoughtfully covered by a patent on 1st July 1920. Two free-mounted contra-rotating rotors were thought likely to provide sufficient lift. Unfortunately in trials it was found that they turned at vastly different speeds thanks to an aerodynamic quirk known as the biplane effect. Worse, and before the problem could be addressed, a combination of gyroscopic effect and periodic unbalanced spinning rotors joined forces with the narrow undercarriage and threw the aircraft onto its side with predictable results.

The experiments continued through numerous rebuilds and fresh designs, many false hopes and quite a few smashes. On 17th January, 1923, his fourth design actually flew at Getafe Aerodrome near Madrid. The pilot was Cavalry Lieutenant Alejandro Gómez Spencer and he made a straight and level controlled flight of 183 metres at a height of about four metres across the airfield. The feat was repeated several days later before civil, military and government officials. During one of these tests the engine suddenly stopped and as the watchers gasped the machine gently settled back onto the ground. Autorotation had been unintentionally proven beyond all doubt!

One would have thought that this might have ensured that the path ahead for Cierva would be smooth, yet it was not to work out that way. His family fortunes had been seriously depleted by the development process. What outside funding was available was sporadic and fell short of the military contract he hoped for. There was indeed some government help, but the truth was nobody could quite see a place for the Autogiro.

Cierva turned first to France for help hoping to set up a co-operative with an established aircraft manufacturer. The French were indecisive: Cierva did not have time on his side so he turned next to Britain, trying to interest first Vickers at Weybridge and then Westland at Yeovil. The choice of Westland was curious for this was a small and impoverished firm at that time and was having ends-meeting problems without taking on foreign development work. On the other hand, Vickers was interested enough to send out a team including its chief designer to Spain to see Cierva and his aircraft. Everything seemed to be going fine until the team got back to Britain and attempted to replicate Cierva's test results in the laboratory. The discovery of a rogue decimal point was blamed for scuppering the whole venture: Vickers concluded that the Autogiro didn't really fly and so they could not be interested.

Around this time, a young American named Harold Pitcairn heard of Cierva's work and was so impressed that he travelled to Spain to meet him. The young American, who had made his fortune in the manufacture of plate glass, found a man who couldn't speak English but was nevertheless charming. One can only imagine the scene of confusion – but Pitcairn did get to see the Autogiro and it made a lasting impression upon him.

Now with a young family to support, Cierva's development work was now proceeding slowly as and when funds allowed but he was not short of good friends and admirers in high places. His latest machine was the C.6 and it was a good flyer. Further improvement produced the C.6bis which was a thoroughly practical machine that handled well.

News of his achievements spread far and wide. And finally the boffins at Farnborough got themselves interested. So much so, in fact, that they invited Cierva and his C.6bis to come and demonstrate at the Royal

Aircraft Establishment's famous Hampshire home. Cierva was no stranger to Great Britain thanks to his father's business connections. Although he could not speak the language, he agreed, he was, after all, fluent in French and most English people, he hoped, could converse with him in that tongue. As his Autogiro was carefully stowed on board a ship for Southampton little could the Spaniard have realised that shortly his life would be transformed and that not only would he learn to fly but he would become his own company's test pilot, he would quickly learn English and that within a very short space of time he would be invited to present a learned paper on his Autogiro before the prestigious Royal Aeronautical Society.

On 4th October 1925 Cierva and his Autogiro (its designation now anglicised to C.6A) were standing on the turf at Farnborough's Laffan's Plain. Six days later, in front of a party including the highest echelons of government and the Air Ministry, hired British test pilot Frank Courtney put the C.6A through its paces in British airspace. The upshot was the agreement that Farnborough would invest some money in further development. In the long term, however, Cierva would be expected to 'paddle his own canoe', form a company to continue development – or go home.

In the end Cierva was to stay in Britain and within a year he was disarmingly fluent in English, able to joke and talk technicalities in his newly-adopted language. Meanwhile development work proceeded with both RAF-marked military models and civil Autogiros with civil registrations. Coincident with this, there was more and more contact with Harold Pitcairn in America. This culminated late in 1928 when Cierva and his C.8 Mk.IV boarded a ship for the United States – and captured the imagination of the Yankee nation. It would be the first of many transatlantic trips by both men to see the other.

AMERICAN INVOLVEMENT

It soon became apparent that Cierva's concept demanded the sort of treatment (meaning expensive investment in design and development) that needed real finance. Development work is never cheap and without government or military commitment, money sources risk rapid evaporation. That kind of money, though, was still to be found in Depression-hit America.

What happened over the next few years would have been impossible had it not been for the total commitment and boundless enthusiasm of Harold Pitcairn who put his personal wealth behind his faith in the Autogiro.

It is estimated that the total spent by the Cierva Company in Great Britain on Autogiro development was £750,000 which, at that time, was worth about $3.5million. Already through the nascent Pitcairn-Cierva Autogiro Company of America (formed in February 1929 as a subsidiary of Pitcairn Aviation, Inc), Pitcairn said that American investment would be no less that $4.5 million.

Two years younger than the Spaniard, Harold Frederick Pitcairn was born at the cost of his mother's life at Bryn Athyn, Pennsylvania, in 1897. Right from an early age he was consumed by an interest in aeronautics and built a glider when he was sixteen: it did not fly. His father, a deeply religious and circumspect man, recognised his son's passion and, uncharacteristically, paid for him to learn to fly properly at the Curtiss Flying School. America, aware of the war in Europe, desperately needed military pilots so Pitcairn was called up for flight training to Army Air Service standards. The Armistice came before he could see service and he was discharged.

From these beginnings, Harold Pitcairn moved into civil aviation initially as a fixed-base operator giving joy-rides and accepting charter work. This was the age of the burgeoning Air Mail routes in America. The most immediate source of money, though, was likely to be from passenger-carrying joy-riding aircraft. So was born the PA-1 'Fleetwing' to carry four passengers and pilot. Pitcairn was now an aircraft manufacturer. His PA-2 took part in the 1926 National Air Races and won both the economical performance category and the hundred miles speed lap race thanks to the clever if simple expedient of a high-speed in-field engine-change. Pitcairn had established a name for himself and the improved PA-3 entered production, ten examples selling for $2,300 apiece.

Like many others of his age, he made his living on post office mail contracts with his PA-5 Mailwing. More to the point he marketed his aeroplanes to other operators: it was the long-awaited financial breakthrough. Now he formed his own airline which was ultimately sold to become the nascent Eastern Air Lines. In the summer of 1929 Pitcairn took his family for a holiday in England. Here they met with Cierva who had just begun trials of his latest machine, the C.8. Later the same week, Cierva's name was headlined in the world's press: he had just flown his machine to Paris – across the English Channel!

In those times, Pennsylvania and, specifically, the region of Philadelphia, was a hot-bed of aviation development and Pitcairn's home was the centre of much that went on. Here was formed the Pitcairn-Cierva Autogiro Company of America, soon simplified to The Autogiro Company of America, a company formed to licence users of Cierva's patents. And Pitcairn was the first licensee. In the fullness of time, virtually all of America's autogiro development was to occur in and around Philadelphia.

Close by in Camden, New Jersey, Wallace W Kellett played as a boy making model aircraft. In 1929, with his brother Roderick G, he would found the Kellett Aircraft Company with as co-directors the brothers Charles Townsend Ludington and Nicholas Saltus Ludington. In charge of engineering was Wynn Laurence LePage. Pitcairn, Kellett and Ludington were already old friends and had met in 1923 as members of the Aero Club of Pennsylvania.

It was logical, then, that after Pitcairn, the next licensee of the Cierva patents should be Kellett. Unlike Pitcairn, the Kelletts had no previous aircraft manufacturing experience although the Ludingtons had pioneered some airline activities.

Initially the Kelletts went their own way, designing the K-1X, a strange machine with a 'fixed wing' carrying differential ailerons and elevators at the rear, and a pylon to support a large two-blade rotor which looked like a full-sized wing. Lacking their own manufacturing capability, the rotor and wing were built for them by Kreider-Reisner Aircraft (later to become Fairchild) at Hagerstown, Maryland, while the fuselage was made by The Budd Company in Philadelphia. Trials began on 14th October, 1930, but with lack of pre-spin the rotor never attained take-off speed. At 58 mph it turned at just 150 rpm and the aircraft showed no sign of becoming airborne. Trials were abandoned on 30th December by which time the Kellett Autogiro Corporation had been established and plans made to obtain a Cierva licence.

HANWORTH ACTIVITIES

With a happy manufacturing arrangement existing between Cierva and the Avro Company, the Cierva Autogiro Company (head offices at Bush House in London, works at Hanworth Air Park in West London) was well-established. By alliances with G & J Weir Ltd in Scotland, Cierva had the services of a sound design and engineering establishment. With James A J Bennett, Cyril George Pullin and James George Weir on the board a proper business structure existed. And as a business it was expected to perform the way all other businesses were expected to perform – to earn money for its investors. It had already opened a flight training school but, relatively speaking, that only brought in beans.

This was all very well, but Cierva, like many before and after him, was always improving things and just when it looked as if a production model might be about to materialise, he'd change it. The C.19 was the first machine to go into volume production – if 34 models of five varieties can be considered to justify the term 'volume'.

Cierva also encouraged other British companies to design and build Autogiros with his assistance. These included Westland, de Havilland, George Parnall and Comper. For Cierva, life was a continuous urge for improvement and while that kept him happy, it exasperated his fellow directors. Matters were already coming to a head when the problem solved itself with the death of Juan de la Cierva in the Croydon air disaster on 9th December 1936. The KLM DC-2 in which he was a passenger crashed on take-off in fog. Cierva was just 41 years old.

The Cierva Autogiro Company continued, by now with the C.30A in production. But Bennett and Weir, clearly aware that the future lay in helicopters, were already moving away from the Autogiro. In America, Harold Pitcairn was horrified at such a move. He had staked everything on the success of Cierva's invention and now his late friend's own company was turning its back upon its founder. As a director of the British Cierva company, Pitcairn sailed across the Atlantic to attend all meetings at Bush House but soon realised that he was no longer one of the other directors' confidants and that, while he gave freely of all developments taking place in America, he was being frozen out of discussions. Distressing as this must have been, Pitcairn vowed to soldier on. Crucially, some time before his death, Cierva had the foresight to realise that all was not well with the Hanworth/Bush House set-up and that perhaps his pioneering Autogiro work was being compromised.

In a curious and only recently revealed deal, he assigned all his valuable patents to Harold Pitcairn. They would earn his quiet American friend a fortune in the years to come – only Pitcairn would not live to see that day.

EMPHASIS SHIFTS TO AMERICA

In 1931, 30-year-old Richard Hickman Prewitt joined the Kellett Autogiro Corporation as designer which he doubled with project engineer to the Autogiro Company of America. In 1933, he became solely a Kellett employee, first as design engineer, then as chief engineer and finally as executive engineer until 1938 when he became vice-president, a position that he retained until 1946. In 1937 John M Miller joined the business as test pilot later replaced by Lou Leavitt. Kellett had by now became the second Cierva licensee obtaining its licence from the Autogiro Company of America. Its first Cierva-licensed machine, the K-2, was a side-by-side two seater having a four-blade rotor and a small fixed wing. This machine launched the company into its most fruitful period during which it went from strength to strength.

Wallace Kellett had met Cierva at Harold Pitcairn's airstrip and now he was 'on board' as an Autogiro-maker he thought he, together with Prewitt, would pay a visit on Cierva in London to talk about the latest developments. Of course, when news of this trip reached Harold Pitcairn he was far from pleased. After all it was he who had cultivated Juan de la Cierva and if there was any visiting to be done, then it should have been organised through him.

This was not the only negative mark chalked up by the Kelletts. While they were as enthusiastic as Pitcairn, they were perhaps just a little naive. At this particular time, the Japanese had a reputation for attending European trade fairs and buying one example of everything on show – then taking the items home, dismantling them and copying them! The Kelletts, no doubt unwittingly, did not realise this when they sold one of their Autogiros to Japan without first negotiating with or through The Autogiro Company of America. This incensed Harold Pitcairn and, coming on top of their private visit to Cierva, created something of a rift between Pitcairn and the brothers Kellett. How justified this discord really was we may now merely guess, but the Japanese had already imported three examples of the Cierva C.19 Mk.IV direct from Hanworth. The arrival of a Kellett K-3, which, unlike the Cierva models which were the first to have three-blade cantilever rotors, had four wire-braced rotors and was thus of an already superseded technology, did not lead to an explosion of Japanese look-alikes.

In fact, no production followed and it was not until 1939 that one of the latest Kellett KD-1A machines was imported. An improved version, the Kayaba Ka-1A direct-control Autogiro was then built in some numbers apparently under a licence agreement with Kellett – another cause for Pitcairn's anger since all licensees were supposed to re-licence through The Autogiro Company of America.

Despite all this, the Kellett business ultimately became more practical in its approach, concentrating on making Autogiros that would suit the needs of the US military authorities. And, when the time seemed right (1943), they seamlessly shifted into helicopters as The Kellett Aircraft Corporation. In this regard they were fairly successful but in the end the bugbear of under-funding caught up with them and, despite a useful order backlog, in 1946 they fell victim to rising costs and inadequate working capital. Having developed what was then the world's largest passenger-carrying helicopter – the XR-10 – and first flown it on 14th June 1947, they were forced to sell this and the even larger XH-17 design to Hughes that August. Bankruptcy took the Kelletts out of what was really no longer even an equation. They successfully broke free of bankruptcy in 1952 but the great days were long past. As a name, Kellett disappeared in 1987.

DORMOY'S CURIOUS PUSHER

After Pitcairn and Kellett, the third licensee for the right to build Autogiros was the Buhl Aircraft Company. Founded in 1925 as the Buhl-Verville Aircraft Division of the Buhl Stamping Company of Detroit, Michigan, it was the creation of Lawrence D Buhl and Alfred Victor Verville, a noted American aircraft designer who was born in 1890. It produced a number of successful light aircraft under the names Airster and Airsedan until 1927 when Verville left to form his own aircraft company. At this point the business was re-styled as Buhl Aircraft Company at Marysville, Michigan. Three years later the company moved to St. Clair, Michigan but in the interim had designed and built the famed Bullpup aircraft, several of which were to compete in the 1931 transcontinental race.

Replacing Verville as chief designer was a French émigré named Étienne Dormoy who is remembered for a rather basic light plane of the 1924 period called the Dormoy Bathtub. Following a protracted watch on the development of the Autogiro in America at the hands of men like Pitcairn, in 1931 he foresaw a market for an Autogiro designed from the outset as a photographic platform. Both civil and military surveillance requirements, postulated Dormoy, could be met with the right design.

The outcome was Buhl A-1 Autogiro, a pusher powered by a 165 hp Continental A-70 radial engine. Comprising a brief pod fuselage to contain pilot, observer and engine, this mounted its tail on tubular steel booms and incorporated a Pitcairn rotor assembly and undercarriage. Piloted by James Johnson, the A-1 was first flown on 15th December 1931.

While Cierva had patented a pusher configuration for an Autogiro with his earliest inventions, Dormoy's design gave the world its first pusher gyroplane – a configuration that has now become the most popular thanks to the amateur end of the market. Buhl's Autogiro was taken on a States-wide sales tour to promote its capabilities. With the encouragement of Cierva and Pitcairn it seemed certain that Buhl would enter production and become a significant player in America. It was not to be for it was the era of the Depression that followed the 1929 Wall Street 'crash'. The Buhl Aircraft Company was severely affected as its aircraft manufacturing business dwindled. By 1933 the business could survive no longer and went under. Only one Buhl Autogiro was ever built. By chance its remains were unearthed in a barn some ten years ago and this unique aircraft has recently been restored as an exhibit in the Hiller Aviation Museum, San Carlos, California.

There was another, fourth, licensee of The Autogiro Company of America and that was a business called F W Steere of White Plains, New York which was granted a license in January 1932. Steere appears also to have become a victim of the worsening Depression for nothing was ever produced by that company.

HOW THE AUTOGIRO FAILED FIRST TIME ROUND

Having said that Wallace Kellett's enterprise was more successful than Pitcairn, such an outcome is oft the lot of the pioneer. And make no mistake, Harold Pitcairn was the American pioneer who devoted his life and capital to perfecting the Autogiro. His was almost a personal crusade to uphold all that Juan de la Cierva had stood for. He abhorred the idea of the helicopter and was not pleased to see the way in which British industrialists James Weir and James Allan Jamieson Bennett were systematically moving the now-founderless Cierva Company into powered rotors. As far as Pitcairn was concerned, both his and the Cierva businesses had agreed a free interchange of information. He at least kept his part of the deal until, realising that it was pointless talking to the British about Autogiros when they were simply not interested, he quietly stopped what was now a one way information route.

But Pitcairn for all his honourable charm was more than unfortunate for he tried to persuade the ordinary American, fresh out of Depression, that the Autogiro was the light plane of the future. The simple fact was that most private aircraft journeys in America were of considerable distance and carried out in ever-faster, more comfortable light aeroplanes. The Autogiro was not able to cruise anywhere near as fast and its short take-off and landing role was only of use at the start and finish of a flight – except, of course, in the case of bad weather emergency. Overwhelmingly, however, the Autogiro was not a comfortable aircraft to fly on long journeys since there was a constant if small oscillating motion in otherwise straight and level flight.

Like Cierva himself, Pitcairn seems to have been in love with the concept rather than in understanding of the market and what it wanted. If Cierva was unfortunate in being the ultimate perfectionist, Pitcairn was the William Micawber of American private flying – Charles Dickens' ever-optimistic character from David Copperfield who always thought that 'something would turn up' if only he pressed on and waited long enough.

With the outbreak of the Second World War in Europe, Pitcairn enthusiastically supplied a fleet of PA-39 Autogiros for the British Armed Services. The consignment was split into two batches travelling on different ships. One ship was torpedoed and went to the bottom taking with it five of the seven Autogiros but, more importantly, the entire complement of spares – rotor blades, engines and so on. The two surviving RAF Pitcairns, devoid of basic maintenance back-up, saw very little service – a sad end to a gallant enterprise.

Soon afterwards, the remaining market for Pitcairn Autogiros faded, the senior US directors experienced a fiery argument and parted – and the company was sold off.

PITCAIRN SUES THE US GOVERNMENT – AND WINS!

During the 1939-45 war (or, as far as the Americans were concerned, the 1942-45 war) great strides were made in helicopter development. These improvements would not have been possible without the Cierva Patents which now belonged to Pitcairn.

The war in Europe, while it did not affect the United States until Pearl Harbor in December 1941, was the catalyst in the development of the Autogiro. Harold Pitcairn gallantly gave his beloved Willow Grove airfield outside Philadelphia to the United States Navy as an air base. When great strides were made in the development of the helicopter, Pitcairn (remember he now held all the Cierva patents) claimed no royalties, anticipating that once hostilities were over, he would be recompensed. This was, as it turned out, wishful thinking. There was particular concern that Bell Helicopters, among others, was capitalising on the Cierva-Pitcairn patents. After the war, The Autogiro Company of America was forced to institute legal proceedings against the US Government for damages.

It was a horrendously expensive and long drawn-out case which progressed from year to year. One only considers a legal challenge of the administration if you have a watertight case – and a good team of expensive lawyers. Harold knew he had the first and made sure he hired the second, but he himself was long dead before, on 25th January 1978, the US Government case was thrown out by the Supreme Court – and a cheque for $31million sent to The Autogiro Company of America. Twenty-eight years of litigation was at an end.

Pitcairn's death came about in a tragic, unexpected manner. On the evening of 20th April 1960, after a convivial evening with family and friends to celebrate his 62nd birthday, Harold Pitcairn bid everybody good night and prepared to lock up for the night. He went into his study. The sound of gunshot rang out and Pitcairn was found slumped across his desk, a pistol before him and a bullet through his right eye. There had been no intimation of personal or business problems, no reason could be found.

Harold Pitcairn's death was attributed to 'accidental suicide'. Unfortunately the newspaper left out the word 'accidental' and the outcome became something of an incubus that stalked the Pitcairn family from thence forward. A recent biography by Carl Gunther included an account of the County Coroner's judgement. Gunther was privileged to examine the weapon which was a small Savage automatic with a hair trigger. Pitcairn always kept the pistol with him when locking up his house because of the threat of intruders. The coroner recorded a reconstruction of what occurred based on the evidence that, on this fateful occasion, two shots were fired, the first probably caused by the vibration of Pitcairn's elbows on the desk: it went through the ceiling. The sound probably startled Pitcairn and his arm jerked, discharging the second and fatal missile into his eye.

Pitcairn, said the coroner, 'was not the sort of man to violate his very deep spiritual convictions which were a life-long practice with him'. Furthermore, determined self-killers would hardly aim for the eye: usually the temple or the mouth. Pitcairn, then, died by his own misfortune, not premeditated suicide. His son, Stephen, upheld his father's memory and achievements to the end. He restored and flew a Pitcairn PA-18 Autogiro which he finally donated to the EAA Museum before falling victim to cancer on 29th March 2008 aged 83 years of age.

Unfortunately the Cierva connection was lost many years earlier in 1936 and with the passing of 'young' Pitcairn, the links to the past, seemingly tantalisingly close at hand, have really receded into history.

A MEMORABLE ERA COMES TO ITS END

The first era of the Autogiro effectively ended with the outbreak of the 1939-45 War. In fact, the end had begun much earlier when Juan de la Cierva's directors in Britain became increasingly exasperated at Cierva's seemingly constant urge to improve his machine. Putting the C.30A into production raised the chances of the company making a profit but still their Spanish chief wanted to improve. The British members of the board of Cierva (which group included Harold Pitcairn, of course) privately believed that helicopters were the way ahead. Pitcairn was never privy to these maverick thoughts but once Cierva had been killed, a new management was rushed into place to expand the work of the Weir company in Scotland which, initially under Cierva's guidance, had been progressing inexorably towards the powered rotor.

It is significant that before he died, Cierva had made over all his patents to The Autogiro Company of America rather than to his British company. This underscores the bond which Cierva had with Harold Pitcairn who persevered with the Autogiro and took Cierva's ideas to new heights in America.

The failure of the Autogiro to capture a large market share was, with hindsight, easy to understand. It was essentially a STOL aircraft and this was its single endearing characteristic. And if you encountered bad weather en route, you could simply stop and land almost anywhere.

The real problem was that the Autogiro was better suited to a country like England where distances were not too great whereas in America journeys between places were much, much greater. This meant that flying times on a trip represented a far greater sector length than comparable trips in Britain. And the Autogiro was not the most comfortable of aircraft to fly for long periods.

Another overriding factor was that the American economy was slow to recover from the 1929 Wall Street Crash and the 1930s Depression. Even after Franklin D Roosevelt's promise of a 'new deal' for the impoverished American public, unemployment continued to climb and firms in all spheres of activity went to the wall. In that time, there were many lightplane manufacturers in the States – and many of them failed because they could not sell enough aircraft, even after repeated price cuts.

The Autogiro was up to three times the price of the average light aircraft which offered a better degree of in-flight comfort. In short, you were being asked to pay for the merits of short take-off and landing. At a time of rising unemployment and starvation for the poor, it was not difficult to see which way things would go. If the world decided it no longer needed Autogiros, then you couldn't force hungry, jobless people to buy one, especially if they'd just sold their cars to buy food.

GYROPLANE REDIVIVUS!

As the first Autogiro era died, helicopters became the 'in' thing not just for the police and rescue services and the military, but for the rich and famous and the inevitable aspirant and unwisely-wealthy show-off. The problem with helicopters was that they were enormously expensive to buy, to maintain and to operate at a period when fuel prices were rising. Also, helicopter piloting required a special licence which was expensive to earn and costly to maintain. While you might easily convert from a fixed-wing licence to an Autogiro licence, there was no equivalent or comparable path to the helicopter.

These were the post-1945 conditions which created the right mix of elements to see a reprise of the autogyro and the first man to make a success of it was an American, Igor Bensen, who designed a rather basic gyroplane powered by a target drone engine.

In the way of things it was not long before Bensen's Gyrocopter reached the shores of Britain where several people recognised the opportunity to improve on it and eventually to evolve the design until it was so far removed from Bensen's original that it entered the realms of a new aircraft. Among the British engineers who embraced the new-era of the autogyro were men such as Ernie Brooks, N D Hamilton-Meikle, Brian Luesley, Peter Lovegrove and Donald Campbell, all of whom contributed greatly.

It was Wing Commander Ken Wallis, though, who had the knowledge combined with the wherewithal to drive the British post-war gyro development programme and his WA series have achieved great things. One of his Wallis Autogyros, a WA-116, gained world-wide fame as a miniature 'autogyro gunship' in a James Bond film.

Not just in Britain has the autogyro been reborn but in Italy, Germany, France, Scandinavia and Japan. The modern form of Cierva's brainchild is actually everywhere but in the sporting or amateur areas of private flying.

It is hard to think now that it's 75 years since Juan de la Cierva's round, smiling face beguiled us. He would, I think, be pleased to see what the world has achieved with his ideas. Above all, he would probably be pleased to know that we – that's you and me and everybody else – have never forgotten him.

Juan de la Cierva, Senior, father of the Autogiro designer, was the one who financed the majority of his son's early experimental work in Spain. A successful lawyer as well as an influential and important local government plenipotentiary during the early trials and tribulations of the Autogiro, by 1931 he had become Minister of Commerce in what was then the newly-formed cabinet of Spain's Conservative leader Juan Bautista Aznar-Cabañas, former Admiral and now Prime Minister following the deposition of King Alfonso XIII. Of course, at that time, Spain's volatile politics meant that nobody in office was safe. Only two months after this picture was taken on 18th February 1931, Spain declared itself a Second Republic and ended the monarchy. Now spearheaded by socialists, Spain was heading inexorably towards civil war. The worsening situation climaxed at the 1936 elections when the Popular Front Party – a left-of-centre organisation – successfully brought together other leftist factions to contrive a poll win. The turmoil was not abated, though, and an army rebellion began on 18th July 1936. This marked the start of the Spanish Civil War and, even though he was now settled in England, the troubles of his homeland and, more particularly, his family caused Juan de la Cierva Junior, great concern. With increasing frequency, he took return trips to Spain, flying from Croydon Airport, to help in whatever way he could. As matters worsened, he planned an urgent return. It was 9th December 1936 – a dull, overcast and quite foggy morning as he boarded a KLM DC-2 airliner, PH-AKL. A few moments later he, along with all the other passengers and crew – fourteen in total – were lying dead in the front garden of a house in Hillcrest Road, South Croydon.

The first Autogiro that Juan de la Cierva made was built in 1920. It was heralded by a Spanish patent application dated 1st July which was granted on 27th August. It was entitled *Nuevo aparato de aviación* – a new type of apparatus for flying. For 25 year-old Cierva it was a monumental year for the previous December he had married and followed his father into local government. The climax of his experiments with model gyroplanes had to be built as cheaply as possible so a French-built Deperdussin monoplane of 1911 vintage was obtained, shorn of its wings, and converted to support a mast carrying a pair of four-bladed contra-rotating rotors, Above that was a fixed vertical fin. The original tail unit, undercarriage and 60 hp Le00 Rhône engine were retained. The test pilot was to be Cierva's new brother-in-law because Cierva himself could not fly. After several runs across Madrid's Getafe Aerodrome it became clear that the machine was unwilling to leave the ground. The rotors turned at different speeds due to the biplane effect; the upper one turned at 110 rpm but the lower one would not go faster than 50 rpm. One final attempt was made but the gyroscopic effect combined with the periodic unbalanced lift and, not helped by the old narrow undercarriage of the Deperdussin, the whole machine rolled over onto its side and was destroyed.

Cierva's second machine, actually retrospectively classified as the C.3, was also built around an old aircraft, this time using the fuselage and other parts of a Sommer Monoplane with a 50 hp Gnôme engine. This time a single three-blade rotor was tried, this time with wide-chord blades. Constructed and tried in the spring of 1921, this, too, was unsuccessful, once again rolling over onto its side and smashing.

The third machine Cierva built, confusingly allocated the reference C.2, was equipped with a 110 hp Le Rhône 9Ja engine and a massive five-blade rotor. This machine was marginally more successful because it achieved several small but encouraging hops but these were at the cost of numerous accidents due to ground resonance which caused the aircraft to tip over on its side without warning. It underwent numerous costly rebuilds but finally trials were abandoned in April of 1922.

After so many problems, all of which involved delay and mounting costs, Cierva built the C.4 which had a well-braced rotor pylon slightly offset to one side in an attempt to counter the tendency for tipping over. Four blades were used in the rotor, each rigidly braced one to the other with very tight cables at the span position. It was also decided to fit ailerons which were carried at some distance from the fuselage not on wings but on braced steel tubular spars. These, it was hoped, would provide a more positive lateral control. On 17th January 1923, Cavalry Lieutenant Alejandro Gómez Spencer made the first ever successful flight in an Autogiro in this machine when he flew straight and steady a distance of 183 metres (600 feet) at a height of about four metres (13 feet) across Getafe Aerodrome. Juan de la Cierva's dream was realized! Well, almost!

A poor snapshot of the historic occasion as Gómez Spencer makes the first ever Autogiro flight in the Cierva C.4 at Getafe Aerodrome. This event was witnessed by a number of important people including the director of Spanish air services and the president of Spain's Royal Aero Club. Within a very short space of time, news of the achievement had traversed the globe. Within days it was being talked about in London and far-off New York. Cierva, however, was running out of finance for his researches and his attempts to get British companies such as Westland and Vickers interested had failed.

Opposite upper: The next venture was the C.5 first flown with a three-blade rotor. Like its predecessor it flew but in July 1923 a rotor blade flew off and the machine crashed. Although it would be some years (and quite a few intervening experimental Autogiros) before it would be adopted for the much later evolution of the direct-control C.30 series, the three-blade rotor was tried out by Cierva as early as the spring of 1923 in his C.5. pictured here at Madrid's Cuatro Vientos Aerodrome. The machine was built around the much-altered fuselage of his earlier and unsuccessful C.2 and construction was the work of Madrid's Industrial College. There was some small government financial aid. Cierva's test pilot (Cierva himself could not fly at this time: he later learned in Britain and became one of his own test pilots) was Calvary Lieutenant Alejandro Gómez Spencer who had made the first proper controlled Autogiro flights in the C.4.

Opposite lower: Flight-testing the C.5 began in April 1923 but it was not as successful as its predecessor being extremely sensitive on the controls. After one minor accident early on which delayed trials, only a few flight; were made before during a take-off run when the aircraft was travelling at some speed, metal fatigue caused one rotor blade to snap off. The resultant massive imbalance tipped the C.5 onto its side and it was destroyed. The pilot escaped uninjured. It was found that the blade had failed due to excessive twisting caused by the massive centre-of-pressure shift on the highly-cambered airfoil section of the blade. Meanwhile some serious research at Spain's government Aeronautical Research Laboratory at Cuatro Vientos Aerodrome was crowned by the amazingly successful flights of a one-tenth scale model. It was then decided that the young engineer Cierva was on to something and finance was offered by the government to build a fresh Autogiro. This time it was created around a redundant British-built Avro 504K, a number of which the authorities had in stock. With fuselage shortened and re-styled as a single-seater, the C.6 with a four-blade rotor proved to be by far the most successful Autogiro to date. Once more news of success spread like wildfire. Cierva improved his C.6, creating the C.6bis or, as it would quickly be known as, the C.6A.

The British Air Ministry invited Cierva to come to England with his curious machine. The Spaniard, knowing that further funding from his family fortunes was no longer possible, and that his own government would be unlikely to fund further development, jumped at the chance to come to Britain. For him it was a great personal leap of faith for he spoke no English, could not fly his own aircraft and, to cap it all, his experienced test pilot from Cuatro Vientos fell ill at the last moment and was too sick to travel. Some while earlier he had met a strange English pilot who had happened over to view his Autogiro at Getafe. They had spoken together in French. Now Cierva remembered his chance encounter and quickly sought him out. An unlikely-looking man to be an ace flyer, this man wore bottle-bottom spectacles and had the appearance of a man of leisure rather than a competent pilot. He was Frank Courtney. Cierva telephoned him – and Courtney became his official test pilot for some while. Cierva's first Autogiro to be flown in Britain was his C.6A constructed using an Avro 504K fuselage and powered by a 110 hp Le Rhône 9Ja rotary engine. This picture, taken at the first press demonstration staged at the Royal Aircraft Establishment (RAE) Farnborough on 19th October 1925. It shows clearly the extended elevators fitted to the Avro tailplane and the ailerons held outboard on a tubular spar. There were no wings and the rotor blades had a wide, parallel chord. Here the C.6A flies at Laffan's Plain, Farnborough.

This close-up of the rotor head displays the manner in which the blades were supported when at rest. Also visible are the pegs protruding beneath the blades around which a rope could be wound for pre-spinning the rotor.

Cierva C.6A as it arrived from Spain with Cierva himself and ready to be demonstrated to a crowd of influential characters at RAE Farnborough by test pilot Frank Courtney. Here he taxies out with rotor spinning. Note the ailerons held out by slender faired spars wire braced to the nose and undercarriage. This picture, taken at Laffan's Plain on Monday, 19th October 1925, shows the short wires suspended from the hub extension to support the blades when at rest. Also visible are the bracing wires which extend between each rotor blade at about the half-span mark to prevent the blades moving towards or away from each other. Aware that some movement was inevitable, Cierva arranged these wires to be slightly slack but with a small centrally-fixed lead weight so that the wires kept taut when the rotor was spinning. Close to the hub on the underside of the blades can be seen the fixed pegs around which a rope was wound for pre-spinning the rotor prior to flight.

While the rotor of the Autogiro would spin up by itself as one taxied along the ground, it necessitated taxiing for a considerable distance before the rotor got up to speed. This was also a dangerous procedure since experience had shown that turning an Autogiro on the ground with a spinning rotor risked rolling it over into an expensive heap. With turning rotor it was best to limit the ground manoeuvring. The solution was to pre-spin the rotor before taxiing. To this end the underside of each rotor blade close to the hub had a projection on its underside. Around this could be wound a rope which could then be pulled by a motor car, a horse or, more usually, a handful of stalwart men. Here at Farnborough in 1925, pilot Frank Courtney gets the rope treatment as he waits, engine running, to start his take-off run.

The Avro 504K-derived Cierva C.6D preparing for a demonstration flight before the press at Farnborough in October 1926. To the left of the picture five men in cloth caps prepare the starting rope which has already been wrapped around the projections under the rotors. A moving-picture camera on a tripod is also visible to film the event. To the right, looking right and wearing his flying helmet, Frank Courtney gestures to a gaggle of trilby-wearing news-hounds while a mechanic at the front drains fuel from the strainer preparatory to starting the engine. The rotor will turn in an anti-clockwise direction – left to right.

Once ensconced at Farnborough, Cierva and the RAE engineers produced a number of Autogiro variations based on the initial C.6A. Here is the C.8V built by Avro in the summer of 1926. It was created using the fuselage of an Avro 552A which was the designation of the Avro 504 trainer when fitted with the Wolseley Viper water-cooled engine. Registered G-EBTX, it was initially fitted with a cable-braced rotor having paddle-shaped blades, stub wings with ailerons and a wide-track undercarriage.

Pictured at a press demonstration in September 1927. G-EBTW, a Cierva C.6D (130 hp Clerget 9Bb) two-seater – the first Autogiro to carry a passenger successfully into the air – overflies the later C.8V, G-EBTX. At this point both aircraft have rotor blades supported by the so-called Farnborough struts or telescopic rotor suspensions. Unlike the original free cables, these struts imposed a very slight resistance to the up and down movement of the blades as they turned which resulted in a rather 'hard' ride. In due course, the struts were abandoned and cables returned. This time, though, they were kept under tension by bungee cords. Still, however, each blade remains secured radially by linking cables at about the half-span position.

There were many varieties of C.6A created under Cierva's instructions at Farnborough, not the least being this C.6L. Notable differences, apart from the undercarriage, include the replacement of the rotor cable suspension by telescopic struts. These 'Farnborough struts' were more or less insisted on by the RAE boffins who were uneasy with these wires which obviously flailed around loosely when the weight of the rotor blade was taken off them once the rotor was turning. However, what they overlooked was that these cables, in being loose, served to allow the blades to twist slightly as they rotated, shifting on each rotation from 'advancing' to 'retreating'. The Farnborough struts introduced resistance into the system with the result that the rotor blades tended to jerk about, making flight tiring and uncomfortable. Eventually these struts were scrapped, but not until the fledgling aerodynamics had proved Cierva to be right and RAE wrong. Later replacement wires featured bungee cords to take up that offensive slack.

Frank Courtney flies the C.6A, now in RAF markings but without any serial number, at Laffan's Plain, Farnborough.

The first ever crossing of the Channel by Autogiro took place on 18th September, 1928, when Juan de la Cierva flew his C.8L Mk.II, G-EBYY, from Croydon to Le Bourget with, as passenger, the editor of the French magazine *L'Aéronautique*, Henri Bouché. Here they are greeted as heroes by public and press alike. That Autogiro is today preserved in the French National Aviation Museum collection.

Cierva was very much the 'man of the moment' and, following his flight across the Channel from London to Paris, the press photographers could not get enough of him. His usually smiling, round face, often shaded by a trilby hat that was several sizes too large for his head, appeared everywhere in newspapers and magazines. Here he was pictured on 20th September 1929 wearing his rather shabby raincoat. Like so many great men, appearance was not high on the list of things to get right.

Late in 1928, the British-built Cierva C.8W was taken to America where it made its first American flight on 19th December. Later it was modified with upturned wing-tips and a revised tail which incorporated a long slender fin. In this state it took the US markings NC418 and the marque number C.8 Mk.IV. This photograph was taken at Dolling Fields, Washington, on 14th May 1929 on the occasion of Harold Pitcairn's flight there from his Willow Grove, Bryn Athyn, airfield to demonstrate the aircraft to US government aviation officials. Today this historic aircraft is preserved in the Smithsonian Institution, Washington.

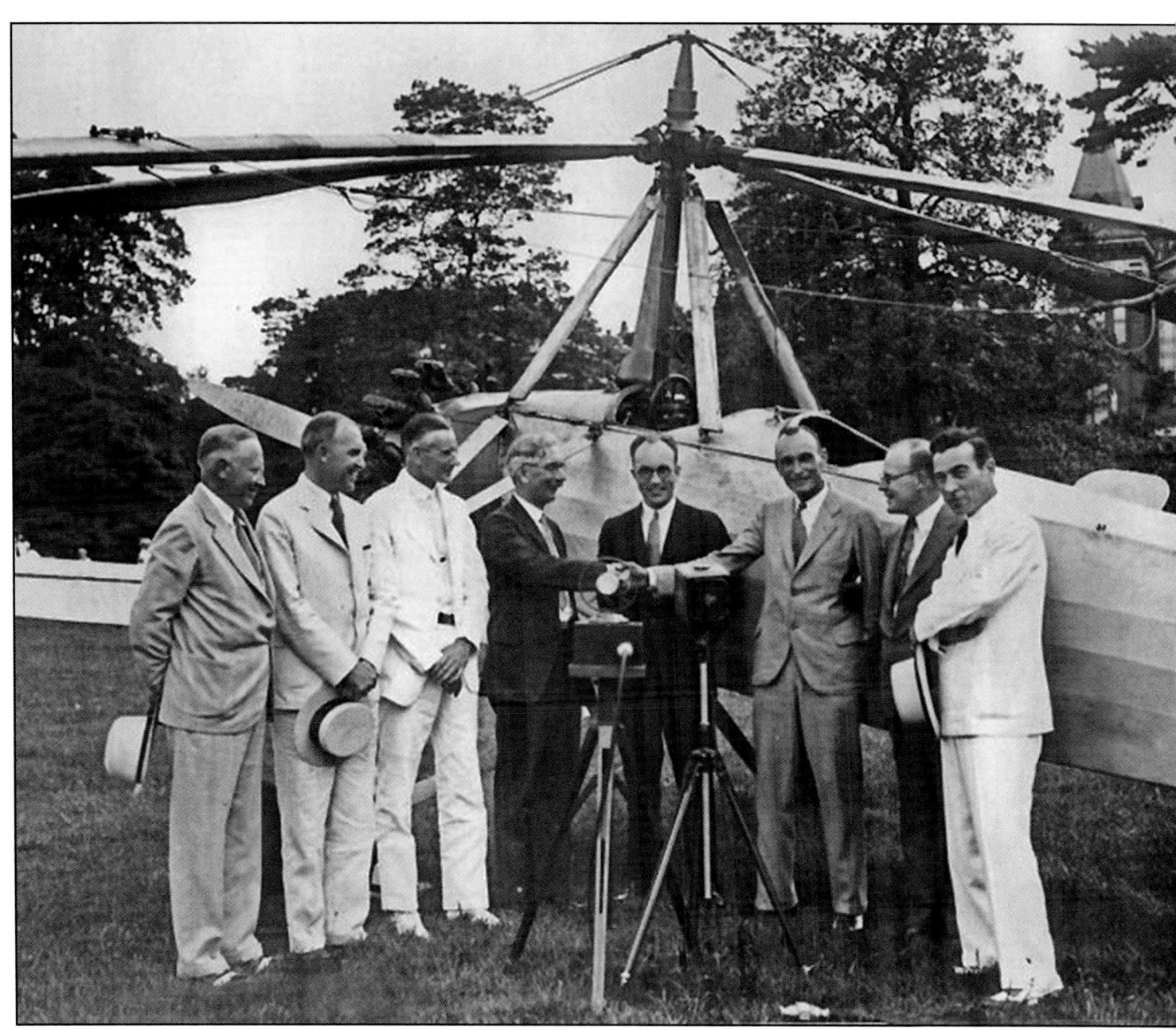

Harold Pitcairn, worried that American-made engines turned in the opposite direction to the British-made Lynx, felt that Cierva's design might not work in the USA. Cierva said rotation was no problem but Pitcairn sent him a Wright Whirlwind to prove matters. The result was the C.8 Mk.IV which Pitcairn himself flew at Hamble in November 1928 before it was crated for shipment. At Pitcairn Field, Cierva pilot Arthur Rawson tested it personally after re-assembly. Now marked NC416, it was the first practical Autogiro to fly in America. Extensively demonstrated, evaluated and modified, it was the confirmation Pitcairn needed to create the Pitcairn-Cierva Autogiro Company of America, Inc, in February 1929. Its working life concluded, it was presented to the Smithsonian Museum, Washington. Pilot Jim Ray flew it to a small park next to the building and here we see the presentation ceremony. Centre is Harold Pitcairn and to his right museum secretary Dr. Charles G Abbott shakes the hand of pilot Jim Ray. Others are, left to right, NACA representative Dr George W Lewis, Senator W R McCroaken Jr. and Smithsonian director Dr. Alexander Wetmore (later to become a renowned ornithologist). Far right we see Geoffrey S Childs, vice-president of The Autogiro Co, and George M Young, assistant secretary of commerce for aviation. Note the heavy tripod-mounted microphones in front of Pitcairn.

Juan de la Cierva quickly learned to fly one of his own Autogiros and immediately took on a major share of the development test flying. Here he lands a C.19 Mk.II, G-AAKY, on the grass at Pitcairn Field, Bryn Athen, Pennsylvania where he demonstrated his machine to Harold Pitcairn on his home territory. This picture clearly shows the restrictive down view from the cockpit and also why Cierva had to cut the lower trailing edge of his fins and rudders to avoid damage in what were invariably tail-first landings. For some reason Cierva eschewed the tailwheel which at this time could have saved much of the problem, preferring the iron skid.

Cierva C.19 Mk.III G-AAYN first flew in July 1930 and the following October was sold to The Autogiro Company of America in the shape of Harold Pitcairn where it assumed the American marks X3Y. This rather poor snapshot shows it on display in the summer of 1932.

This C.19 Mk.II, G-AALA, got its Certificate of Airworthiness on Christmas Eve, 1929, and was involved in development work including being lent to the Air Ministry during March-April 1930. It was flown by a variety of pilots including two Service test pilots. It was found that the deflector tail, also known as the 'scorpion' tail, could spin the rotor to about 50% of take-off rpm necessitating a take-off run of 600 feet in an 8 mph wind. The landing distance was sixty feet. The speed range was from 31 mph to 82 mph with a rate of climb of 540 ft/min. Upgraded to C.19 Mk.III status by the fitting of an Armstrong Siddeley Genet Major II and an enlarged 35 feet diameter rotor, it became the first true production Autogiro. Here it is pictured at Hal Far, Malta, during a European tour in 1931-32. Hal Far was the old aerodrome in the south of the island which was also known as RN Base 'HMS Falcon' opened 1929 and closed in 1979. G-AALA was later fitted experimentally with a three-blade fully-cantilevered rotor becoming a C.19 Mk.IV but crashed in May 1932.

More Cobham displays, this one in Norfolk in 1932. Unlike modern air displays today, in those times once you had paid your sixpence to get into the field you were free to roam at will amongst the aeroplanes since there were usually sufficient staff on hand to deter the heavy handed (or footed) small boy or over-curious adult. Here G-ABGB the Cierva C.19 Mk.IVP is parked close to Avro 504K G-AAAF.

The C.19 represented a major break-through for the Cierva Company. It was the first Autogiro to be designed from inception as a gyroplane. All previous models had been based on adaptations of existing aircraft and built around, predominantly, Avro 504K/N fuselage assembled including undercarriage, engine and tailplane. The C.19 had a dedicated Autogiro-designed fuselage made of welded steel tube and was built by Avro at Hamble. Various models of C.19 were built but common to all was a special tail assembly having both tailplane and elevator separated by a gap so as to create a 'letter-box' structure. The rotor still had four wire-braced blades each with its static-support 'landing wire' bracings. Early trials disclosed that at steep rates of descent (in which condition the nose is held high), the first point of contact with the ground was not the tailskid but the bottom corners of the rudders. This necessitated the sharp up-cut trim to the rear of the fin/rudder assembly. An interim machine between the Mk.II and the production-series Mk.III was the sole example of the C.19 Mk.IIA, the subject of this set of pictures all taken by the magazine *Flight*. Registered G-AAUA it first flew in January, 1930, gaining its Certificate of Airworthiness the following August.

Powered by a 105 hp Armstrong Siddeley Genet Major I five-cylinder radial engine, the C.19 Mk.IIA incorporated a long-travel undercarriage and the then-usual very narrow-tread Palmer low-drag, high-pressure aerowheels. The rotor diameter had been increased from the earlier 30 feet to 35 feet with blades having reduced-chord roots and a tubular steel spar having wooden ribs at three-inch spacing. The blades were still cable-braced.

This head-on view shows the short-span wings with upturned tips. While this tip form served to increased the stability at low speed through the overall increased dihedral angle, it was also an attempt to help minimize damage when conditions of ground resonance arose and the aircraft began to 'dance' from one wheel to the other until, unless quickly corrected, the aircraft stood on a wing-tip, turned over and the rotor blades dug really big (and expensive) divots.

This view reveals the rotor suspension and also the four inter-blade wires that joined one blade to the next at about the 40% blade length. The little 'spire' on top of the rotor to which the rotor blade landing wires are attached is also a small oil tank to lubricate the rotor upper bearings. Note also the normal setting of the 'letter-box' tailplane for level flight. This model was still some way from the era of rotor pre-spin power take-off drives from the engine.

An innovation with this model was the ability to raise both tailplane and elevator surfaces to form one large upwards-point surface between the fins to direct the engine-driven propeller slipstream upwards into the rotor. This set it spinning and helped reduce the length of time needed for the rotor to speed up sufficiently to enable a safe take-off to be made.

Demonstrated to the aeronautical press at Hamble, the Autogiro was parked immediately in front of one of the large hangars, began its take-off run and managed to climb out over the building. In this picture, with rotor spinning and the elevator alone in the up position, the C.19 Mk.IIA heads, apparently, for unavoidable disaster.

The Autogiro in level flights clearly showed the steep upcut of the fin/rudder rear lower corner and, incidentally, the slight downthrust of the engine to aid the deflection of the slipstream into the rotor. The optimum rotor speed for take-off was 180 rpm.

The extent of the undercarriage shock absorption is revealed by the amount of undercarriage extension in this fly-by picture. Just visible are the cables that tie each rotor blade to its neighbour at about the 40% radius. These allowed the blades to advance and retard a few degrees during revolution.

The box tail on the C.19 proved particularly demanding to stress, in particular in the condition with both surfaces raised for rotor spin-up. To air this, the rear of the fuselage was quite deep to allow the attachment of tubular steel cross struts at each side. Those on the port side are clearly seen in this view.

In this steep banked turn towards the camera the geometry of the wing bracing is seen, and the rotor bracing wires are just visible between the blades. The fixed wing spanned 17 feet 8 inches and the maximum speed was between 82 and 95 mph. Service ceiling was 8,000 feet and, by cruising at 70mph and the range was 300 miles.

A detail view of the rotor hub showing, from top downwards, the turnbuckles attaching the blade suspension or 'landing' wires carried on top of the conically-shaped rotor-hub lubrication oil tank. The blades are attached first by a horizontal hinge so they can move up and down, and second by the vertical lag hinge which permits to blades to move laterally a small amount during revolution. Just visible between the legs of the attachment pylon can be seen the flexible drive of the rotor speed indicator which was a simple tachometer on the instrument panel that indicated how fast the rotor was turning.

In 1932, Sir Alan Cobham's National Air Days Displays, popularly known as Cobham's Circus, acquired an Autogiro for its fleet. G-ABGB was a C-19 IVP registered on 11th April 1930. On 10th September the Circus came to Edinburgh where, on Silverknowes Farm, some three miles northwest of the city, it set up its flying display and joy-riding business. Here the aircraft is seen parked in the field with blade-tip tether socks in place to prevent the rotor turning in the wind. This was the last year of both Silverknowes Farm and the autogiro. The Cierva went on demonstration to South Africa where it crashed at Cape Town on 17th February 1933, and by the following autumn this fine field was converted into a housing estate, albeit on an imaginative ground-plan.

In flight, the C.19 presented a curious sight with its long, slender wings and relatively large spade-shaped tail.

The young always turned out with glee when the Circus arrived in town. With Sir Alan Cobham's National Air Day displays, the event was popularly referred to as 'Cobham's Flying Circus' and for this touring show both young and old flocked to the flying field with shared high spirits. It was a chance to get up close to aeroplanes, to see brave, daredevil pilots and the opportunity to see equally daredevil (and often dangerous) flying. In those days, before the invention of Health & Safety and aviation display rules, pilots would dive straight at the crowd and perform aerobatics well below a hundred feet. Here the display field is already well-packed as the Circus arrived in formation – a sight and sound to savour as the joy-riding airliners and aerobatic aircraft fly low overhead. At this display the show was joined by G-ABFZ, a Cierva C.19 Mk.IV.

The Cierva C.19 Mk.V was the world's first direct-control Autogiro and was funded by Cierva himself. The prototype was G-ABXP, converted by Avro from the last production C.19 Mk.IV built at Hamble in the latter part of 1931. Initially the tilting head was worked by push-rods. Meanwhile, the Spanish Air Force bought a C.19 Mk.IV in December 1932. This was G-ABXI which became EC-W13. The Spanish now took a great interest in the direct-control developments at Hamble and Hanworth and consequently made their own conversion of the Autogiro into a C.19 Mk.V status, first as EC-ATT, then EC-CAB and, latterly, as EC-AIM by which time the one-time G-ABXI was a true hybrid – a winged model C.19 fitted with (by then) a C.30A three-blade cantilever rotor assembly and a straight-cut top to a non-standard fin and rudder. There is a plain, simple, straight tailplane and elevator. Its ultimate fate is unknown. This picture is from around 1936.

Novel amongst the Cierva Autogiros at Hanworth was this one, G-ABXP. Built by Avro at Hamble, it was the direct control development aircraft created without wings, with tilting rotor here operated by a pair of black-coloured push rods, and shoulder-mounted undercarriage. The presence of the extremely low tailplane, single rudder and large tailwheel shows that while this is of the C.19 family, it is the sole Mk.V. Here seen landing after a demonstration flight by Juan de la Cierva himself, the first flight took place in March of 1932. The aircraft was finally scrapped in 1935.

The three-legged rotor pylon and shoulder undercarriage define G-ACFI as the prototype C.30 Autogiro built at Hanworth by National Flying Services Ltd. Its first flight took place in April 1933, an event pictured here and conducted by Cierva himself. The C.30 led directly to the four-strong C.30P preproduction batch that led to the production model, the C.30A.

The Heart of the Autogiro

as manufactured for the latest type
DIRECT CONTROL C.30
to the order of the Cierva Autogiro Co.
was machined from bar and assembled by—

The MOLLART ENGINEERING CO.
Thames Ditton, Surrey.

A. J. MOLLART, M.I.P.E.
J. C. HENDRA, M.I.P.E.
Telegram } Emberbrook 1846.
Telephone

Makers of Limit Gauges, Jigs, Fixtures, Press Tools, Screw Gauges.

The hub of Cierva's Autogiro was a highly specialised piece of engineering and manufacture of this was subcontracted to specialist gearbox makers Mollart in Surrey. This company advertised their involvement with Cierva in *Flight* and *The Aeroplane* in the early 1930s. The illustration shows to good effect the design of the C.30A rotor hub, the clutch and drive shaft (with universal coupling) to pre-spin the rotor for take-off, and the blade hinges as well as the hub tilt bearing.

Here is the prototype of the Cierva C.30, an interim machine that presaged the C.30A production Autogiro. G-ACFI was built by National Flying Services Ltd at Hanworth and made its first flight in April 1933. This was the first proper direct control model developed from the C.19 Mk.IV. Its salient points are the three-strut rotor pylon, the fully cantilever three-blade rotor, the pronounced 'elephant's trunk' control column curving down into the rear cockpit, and the haunch or shoulder-type undercarriage soon to be replaced by the cantilever-truss style. After the C.19 series' tailskid, a proper large-diameter pneumatic tailwheel (suggested and implemented at the suggestion of Harold Pitcairn) was fitted. Very much a development airframe for the C.30A, G-ACFI was scrapped in 1938. In this photograph the passenger in the front seat is trailing a long cord to something in the grass out of shot.

Cierva Autogiro Company at Hanworth operated a flying training school to convert ordinary 'A' Licence pilots to fly Autogiros. In 1935 it also offered to teach a person to fly from scratch on an Autogiro. Logical thinking might suggest that this would be an unnecessarily restrictive qualification to possess, converting a fixed-wing pilot to rotary wings being a more useful (and cheaper) solution to attempting it the other way round.

NOW!
YOUR "A" LICENCE TRAINING
complete course **£35**
and on the   No Extras
AUTOGIRO, Too!

Here is your opportunity to learn to fly under expert tuition on the latest type Aircraft at a cost that seems almost too good to be true! The course includes all dual instruction and solo flying for Air Ministry requirements.

If you hold an "A" licence you can have dual instruction and one hour's solo for **£5** inclusive.

Full particulars from:

THE CIERVA AUTOGIRO

THE CIERVA AUTOGIRO CO. LTD.
BUSH HOUSE, ALDWYCH, W.C.2.
Phone: Temple Bar 2561.

Cierva C.30A, G-ACUU, was used by Air Service Training Ltd at Hamble. Now fitted with cantilever truss undercarriage and without a rudder, this dated from 14th September 1934. The shape of the rotor hub fairing was finalized. Note the almost vertical slender rotor pre-spin shaft between the front pair of rotor hub pylon struts.

On the way to perfecting direct control, Cierva experimented widely and produced some hybrid aircraft such as this C.30 fitted with Cierva's 'Autodynamic' rotor hub which allowed the aircraft to jump into the air from a standing start after the rotor had been pre-spun by power take-off from the engine. Experimentally fitted with a two-blade rotor it was test-flown by H A Marsh at Hanworth on 23rd July 1936 on which occasion this photograph was taken. The non-standard undercarriage with its unusual bracing is evident as is the uncowled rotor hub and the large area of rudder fitted. Cierva's special rotor head seen here occupied much time and money in development yet in the end was abandoned for it could only be used with a two-blade rotor.

A Cierva C.30A showing the differences between the C.30 prototype, G-ACFI, and the production model. G-ACYH is pictured at a Cobham National Air Displays venue and this photograph reveals the wide stance of the cantilever truss undercarriage and also the four-leg rotor pylon. This model also dispensed with the need for a rudder, the old shape of the fin-rudder assembly being preserved but the movable surface being merely vestigial rather than controllable. Later a small trim-tab was fitted to the back edge of what was now simply one big fin: this was to trim out the remaining fraction of engine-torque-induced turn and was useful to make long distance flight less of a chore.

The AUTOGIRO

The Utility Aircraft.

An **entirely new** type of Light Autogiro, comprising a device for starting the rotor blades, will be on exhibition at

STAND No. 20

at the

INTERNATIONAL AERO EXHIBITION

OLYMPIA

July 16th to 27th.

THE CIERVA AUTOGIRO Co., Ltd.
Bush House,
ALDWYCH, LONDON, W.C.2.

The International Aero Exhibition at Olympia in 1929 was the first chance many had to see a machine close-up and this notice published in *Flight* for July 16th drew many inquisitive visitors to the Cierva stand.

A standard production type C.30A shows off its cranked-tip tailplane and strut-braced undercarriage trusses in this almost head-on view.

39

Accidents to aircraft can sometimes reveal more than merely a broken aircraft and occasionally lift the veil off some inherent design deficiency. One problem with the Autogiro was that the normal technique for landing was to hold the nose high so that the machine might land with as little forward speed as possible. With winged Autogiros this flight attitude severely limited the ability of the pilot to see his touch-down point. Even in the wingless C.30A, looking down was not easy on this sort of approach and unless you constantly dodged from one side of the cockpit to the other, something could easily stray undetected into your path. This is just what happened in this unusual accident at Hanworth between Cirrus-engined BA Swallow G-AFHK and Cierva C.30A G-ACWR. The Swallow had gained its Certificate of Airworthiness (C of A) on 1st June 1938 so the mishap must have been between then and the outbreak of war. There is no information about this clearly major incident but it is easy to reconstruct the cause from the damage sustained. The whole front of the Swallow has been knocked off to the starboard, the engine being upside down against the wing leading edge. The Autogiro's rotors are undamaged but the metal airscrew has been damaged and the engine pushed back, downwards and to the port. During this steep angle of descent of his Autogiro, the pilot would normally look out of his port side. It seems likely that the Swallow, hopefully with only one person aboard (rear cockpit) taxied into the path of the Autogiro as it landed. The huge quantity of oil on the Swallow's starboard wing shows that the Autogiro's engine was still running when its airscrew sliced open the Swallow's oil-tank.

Opposite: A second view of the Autogiro/Swallow mishap showing the rocker-box covers of the upside-down Cirrus engine visible over the wing leading edge by the top of the letter 'H' in the registration. Despite this extensive damage, G-AFHK was rebuilt: see page 133 of *British Light Aeroplanes* (Ord-Hume, GMS, 2000) for a picture of a Klemm Swallow being rebuilt after what appears to be very similar damage. This particular Swallow was finally scrapped at Hanworth in December 1946. As for the C.30A, it had gained its C of A on 1st August 1934, survived to be impressed as V1186 but was eventually lost at sea off Worthing, Sussex, on 24th October 1943.

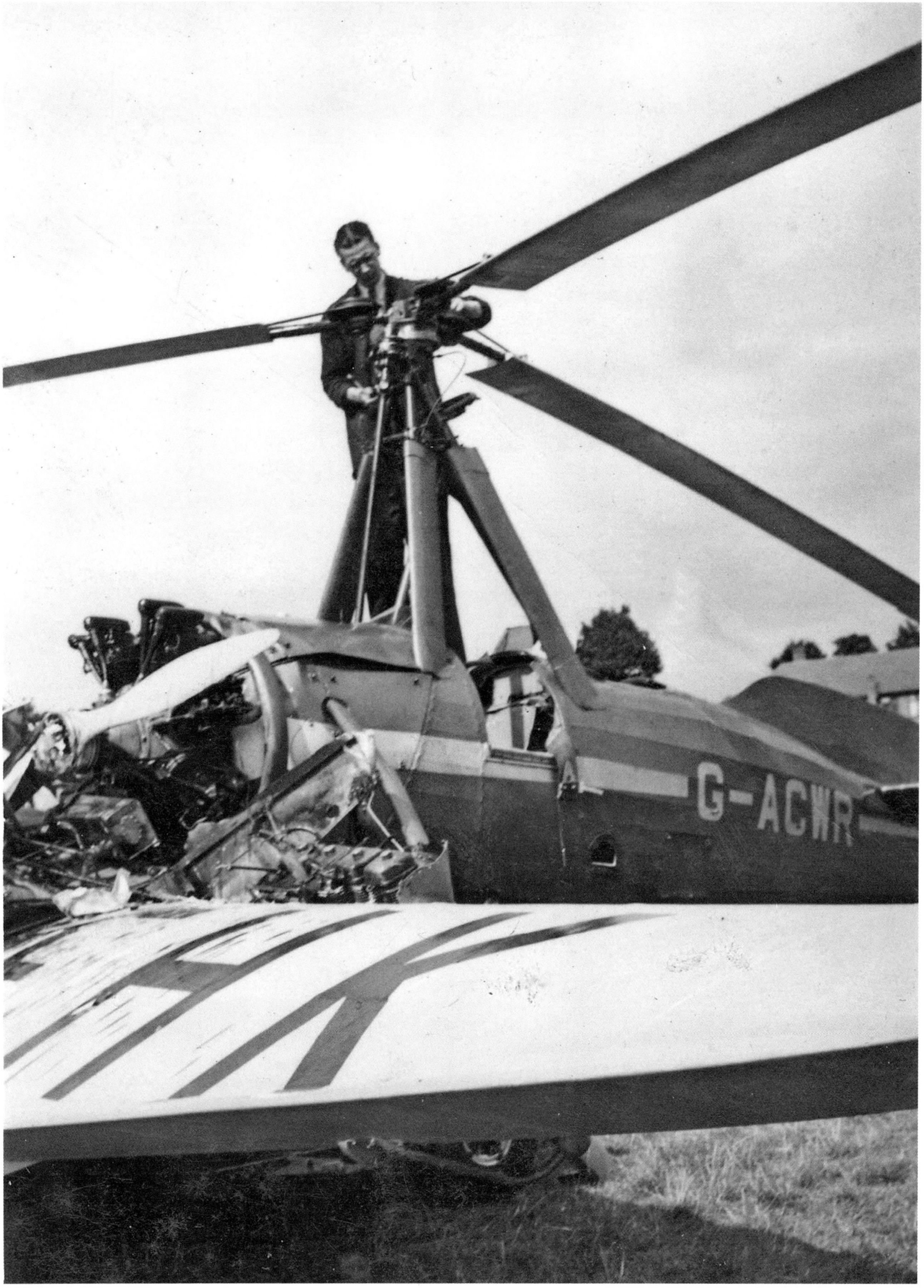
G-ACWR

Cierva C.30A G-AHTZ began life back in July 1934 as G-ACUI, one of the first Avro-built machines. It was first owned by the Hon A E Guinness who later would have a pair of Pitcairns. The Heston-based machine was impressed in September 1942 as HM581. Surviving the war but without its paperwork it was re-registered G-AHTZ by Essex Aero Ltd at Gravesend. The RAF's name for the C.30A, Rota, appealed to a popular make of rotary hand-towels found in lavatories up and down the land; in July 1947 'HTZ became the property of Rota Towels Ltd and, based at Elmdon, was given the name *Billy Boy*. It became a popular star of many a post-war air show and a generation too young to have enjoyed the Ciervas of old marveled afresh at the silver and black Autogiro's antics.

Billy Boy in flight showing off its attributes. Unfortunately, all good things have to come to an end and sometimes the end is terminal rather than graceful retirement. On 4th March 1958, this revamped veteran of the skies crashed at Elmdon and was consumed by fire.

VH-USQ began life at Hamble as G-ACXP, gaining its C of A on 28th September 1934. It subsequently went to Australia where it became VH-USQ pictured here. Keith Gatenby flew this extensively from Mascot, NSW, but it was withdrawn from use in July 1940.

In Milan, the Muzeo Nazionale della Scienza e della Tecnica 'Leonardo da Vinci' at Via Vittore 21, preserves an interesting Cierva C.30A, G-ACXA which dates from 8th May 1935. That August it was sold to Italy becoming I-CIER. Painted cream and red with black registration, it is displayed in this rather congested exhibition but, unlike many such rotary-winged aircraft in preservation, has all its rotor blades in place.

As early as the mid-1930s the first C.30A Autogiros were being drafted for service use, in this case by the School of Army Co-operation at Old Sarum. The first to arrive was K4230, seen here. It had been delivered to the Aeroplane and Armament Experimental Establishment on 5th February 1935. Called, by the services, the Avro Rota I, this was quickly modified to undertake experimental deck landing trials at Gosport after which it was taken on board HMS *Courageous* on 11th July 1935. It later went to HMS *Furious* before serving with No.2 Squadron in 1937. It was struck off charge on 6th March 1939 having amassed just 126 hours and 45 minutes in the air.

The RAF's batch of Rota I Autogiros included K4230 seen here parked on the grass. Interesting from this view is the wide track landing gear and the lack of a rudder, merely a trailing-edge tab being fitted to its enormous fin.

The Cierva C.30A in military guise was known as the Avro Type 671 Rota Mk.I and K4235 was the sixth of ten delivered from A V Roe's Hamble factory between August 1934 and the following May. It is pictured here with the Squadron Code KX-B while serving with 529 Sqdn. For some reason it has had its hub fairing removed. This machine had a chequered service career before being sold to Fairey Aviation Company on 23rd May 1946. It was registered G-AHMJ and was broken up a year later.

The first use of Cierva Autogiros by the Royal Air Force was in the summer of 1936. Here K4233, described as an Avro 671 Rota I, serves as a backdrop for a young airman's portrait taken at Tangmere in that August. The RAF, confronted by a high-risk situation associated with split trousers and other related damage, decided early on that mounting the Cierva, like performing a similar task with a large horse, was best conducted with the services of a stable-boy and/or a box. This snapshot conveniently shows the mounting-box in position.

Cierva C-30A G-ACWH was one of three machines impressed into the RAF. This one became DR623 of Calibration Flight 74 based at Hethel in Norfolk during 1943. The others were G-ACYH (DR622) and G-ACWF (DR624). All three were to survive the war and were sold in April 1946 whereupon they adopted new registrations. DR622 became G-AHRP, DR623 (pictured here with Squadron code letters KX-M) G-AHLE, and DR 624 G-AHMI. These Autogiros fulfilled a unique requirement in assisting in the calibration of radar installations, duties suited to their ability to fly an accurate heading both slowly and repeatably. The airfield at Hethel was, at this time, shared with B-24 Liberator bombers operated by the USAF's 2nd Combat Bombardment Wing, part of 389 Group.

During the 1939-45 war, the Rota I played a vital part in the calibration of our radar defences, invented in 1937. With its ability to fly slowly and hold a precise heading at a predetermined altitude – and to repeat such an exercise over and over again with the minimum of delay – the Autogiro was employed in setting up our costal chain of radars which were vital to detect in-coming enemy aircraft. This rare photograph shows one of these machines and displays its array of calibration aerials supported on a pylon to the starboard side of the rotor pylon and on the compression strut of the starboard undercarriage truss. Both these extend to a folding pylon mounted above the centre of the tailplane: this was normally only raised when the Autogiro was in flight and the rotor blades coned upwards.

A fine study of the Cierva C.40 designed after Juan de la Cierva's untimely death by Otto Reder. Side by side seating in a semi-enclosed cockpit surmounted by an inverted dustbin-type cowl combined with a revised, high-energy shoulder-mounted undercarriage to make this suitable as a military workhorse. Notice also how the rudder had returned to favour in a big way. L7589 was the first of a batch of five Rota 2 aircraft built at Eastleigh by rotor blade specialists Oddie, Bradbury & Cull Ltd. Intended for the Royal Navy they were instead transferred to the Royal Air Force and served with 81 Sqdn. This example was flown to France on 21st October 1939 returning later that month but it did not have a long life, crashing at Odiham on 29th April 1940. Notable here in this fine photograph by Charles E Brown is the large rudder.

Juan de la Cierva was not alone in autogyro development work and one man who operated quite independently from the Autogiro world based at Hanworth was Scottish engineer David Kay. He built two machines that he named 'Gyroplanes' and, technically speaking, they were more advanced than Cierva's models. Incorporating a four-blade cantilever rotor that combined side to side rotor tilting with collective pitch control, the Kay had the makings of a very practical machine. It would be several years before Cierva introduced such features. G-ACVA, powered by a Pobjoy R geared radial of 75 hp and driving a small-diameter propeller having four wide-chord blades, was styled the Type 33/1 and, unlike his first machine which was constructed in a local garage, was built by rotor blade specialists Oddie, Bradbury & Cull Ltd at Eastleigh. It first flew on 18th February 1935 and, like its small predecessor the Type 32/1, it was demonstrated to the Royal Air Force at Leuchars. *Photograph by G Witt.*

When first it flew in 1935, David Kay's Gyroplane, G-ACVR, was technically ahead of the Cierva Autogiro thanks to its use of collective pitch control. Now in preservation in Scotland it is seen here at a 1960s air show at Perth. The huge blades of the four-blade propeller were necessary to absorb the 75 hp geared Pobjoy radial engine's power in such a small-diameter airscrew. In the summer of 1939 a company was formed to acquire the rights from Kay and put the Gyroplane into production: War's outbreak scuppered the project but the prototype aircraft survived and is now in preservation.

Each of the manufacturers invited to participate in Autogiro development in Britain applied themselves to their tasks in quite different ways. Avro's C.17, for example, was an Avian fuselage suitably modified – and later restored as a normal Avian. And when de Havilland had the chance to create the C.24 it emerged as a cabin two-seater with more than a touch of Puss Moth about it. Like so many of its contemporaries, it was below performance expectations and the first flight at Hanworth on 11th November 1931 was a bit of an anticlimax. However, it did inspire the vital newspaper coverage and the press went to town on the fact that this was the first tricycle-undercarriaged Autogiro and that it took off in just 15 yards (thanks to the rotor pre-spin gear now becoming standard with 'gyro hubs'). The revelation that the performance was down and the top design speed could not be achieved was compounded when it was revealed that the C.24 had been designed as a three-seater! The aircraft was presented to The Science Museum which duly looked after it until, during 1974, G-ABLM was fully restored to (almost) flying condition by de Havilland Technical College students.

De Havilland's contribution to the Autogiro's development, the Cierva C.24, was an elegant cabin two-seater possessed of a tricycle undercarriage, and a fine appearance. Originally it had a simple tail like the true Ciervas of the era, but, as with the contemporary Comper, it was quickly found that vertical surface area was deficient, hence the addition first of a dorsal extension to the fin, and then to the tailplane end-plates or 'Zulu shields'. Despite its restricted flight envelope, G-ABLM was nevertheless a good flyer. It was noted, however, that like a racehorse, it was best served by small(ish) pilots since access to the cabin was far from straightforward, entry and egress obstructed by the wing strut and the overhead rotor. Although the engine was removed when the machine went into long-term storage in the Science Museum before the war, as part of its recent display a real engine has been replaced beneath the cowlings.

Closely associated with Cierva was the Scottish firm of J G Weir, James Weir built his first machine, the W.1, at Cathcart in the closing months of 1932. It was designed by a consortium comprising Cierva and Weir's engineers F L Hodges and R F Bowyer. Flights were undertaken at Hanworth by Cierva himself and Alan Marsh. It suffered from excessive vibration and several other problems which were successfully overcome by this, the W.2 (above), which was first flown at Hanworth in mid-1934. Power was provided by a special flat-twin engine designed by Cyril George Pullin and Michael Walker and built by Weir. Called the Douglas Dryad, this was geared and produced between 45 and 50 hp. Although never put into production, this one-off machine was projected as a private aircraft to be offered on the market at £355. It underwent several design changes: originally it had a tailplane with cranked tips like the C.30 but later this was modified as seen here to a straight tailplane with large tip plates. After that the W-3 (pictured on the facing page) was even more advanced for it was a 'jump-start' Autogiro where the pitch of the spun-up rotor was increased sharply whereupon the energy stored in the fast-turning blades allowed the aircraft to leap to a height of about sixty feet. This was the test vehicle for Cierva's so-called 'autodynamic' rotor head which could only be made to work with a two-blade rotor and was quickly abandoned. Power for the W-3 came from an amazing 50 hp four-cylinder inverted inline designed by Pullin and Walker and called the Weir Pixie. The only other aircraft to have this engine was the ill-fated Taylor Experimental monoplane, G-AEPX, built by Richard Taylor who was killed in it on its maiden flight. The first flight was by Cierva test-pilot Henry Alan Marsh at Abbotsinch Airport, Glasgow, on 9th July 1936 before being trucked to Hanworth for official trials. Next came an improved version, the W-4 of December 1937. Plagued with rotor imbalance and ground resonance, it overturned during taxi trials and as a consequence Weir dropped all further Autogiro work and moved into helicopters. Today just the W.2 survives in preservation.

Opposite: The Weir W-3 jump-start Autogiro was the third of the experimental machines made by J G Weir Ltd at Cathcart. It was also the machine about which least is known for it was only flown a few times. It seems that previous experience with ground resonance had influenced the undercarriage design for the unregistered W-3, although quite small, featured a very wide-track landing gear. At 14 feet 4 inches, it was shorter even than the W-l and had a maximum weight of 650 lbs. The top speed was said to be 100 mph.

51

Another Autogiro maker was Comper Aircraft Co, makers of the Swift single-seat racer. Designer Nick Comper used many Swift parts in the realization of the C.25. The smallest and fastest Autogiro of its time, the Pobjoy-powered machine was registered G-ABTO and made its first flight in the hands of Cierva himself at Hooton Park, Cheshire, on 20th March 1932. There were numerous problems, predominantly direction-controllability which contributed to a nasty crash. Rebuilt, Nick Comper himself flew it from Hooton Park to Heston and then to Hanworth before handing it over to Cierva pilot Arthur Rawson. It underwent further trial flights but still fell short of stability needs. After another accident at Heston Air Park in 1933 this elegant little Autogiro was scrapped. *Picture via Richard T Riding.*

Built in France, the Weymann-Lepère C.18 was a two to four-seat Autogiro built under Cierva licence to the design of Cierva and Georges Lepère for Loel Guinness as an entry to the Guggenheim Safe Aircraft Competition in 1929. Novel features included all-metal stressed-skin construction and a landing gear incorporating Messier shock-absorbing struts. It was also the first cabin Autogiro. Registered G-AAIH on 5th June 1929, it went to Pitcairn's Willow Vale airfield where it was refitted with a 225hp Wright Whirlwind to replace the Salmson radial, itself a replacement for the original upright Renault inline engine. High vibration and poor performance directed that it did little flying and soon disappeared.

Above: One of the other Cierva Autogiro builders was Westland Aircraft at Yeovil and it was here that the unfortunate CL-20 was built. Transported to Hanworth by road for testing, it made its first on 5th February 1935 with Cierva himself at the controls. G-ACYI was not a great success and after a few further test flights by Alan Marsh, it was scrapped. The Pobjoy Niagara-powered two-seater had a direct-control rotor assembly supplied by Cierva but, with a passenger on board, the machine would not climb more than a couple of hundred feet: built rather like a battle-cruiser it was both overweight and grossly underpowered. The last flight was on 25th July 1935 after which it was broken up. It was an odd aircraft in a number of ways, particularly regarding the tail. Each of the outer vertical fins, unusually, was of airfoil section with the flat side innermost. Halfway through trials these were reversed. Also the starboard half of the tailplane had positive fixed incidence: the port side was negative! It was an exercise in asymmetry worthy of Blohm und Voss on a good day! The total flight time amassed was eight hours and thirty-three minutes.

When, in the summer of 1929, Juan de la Cierva took his latest Autogiro across the Atlantic to show Harold Pitcairn, he took the opportunity to demonstrate his machine at the National Air Races, Cleveland, Ohio. Here, before that event, he poses with C.19 Mk.II, G-AAKY, at Bethayres Field, Philadelphia. It is 23rd August 1929.

Harold Pitcairn was an established aircraft designer but had long been fascinated by the idea of the Autogiro, so when he heard that a Spaniard had made one fly he immediately made contact with Cierva. At that time, Cierva spoke no English, and Pitcairn no Spanish. Despite such apparently insurmountable handicaps, the two men were to become the very firmest of friends and, after Cierva's early death, Pitcairn continued with the development work Cierva had begun. Having announced he would like to represent Cierva in the US and to have the agency for the Autogiro there, Pitcairn and Cierva jointly formed The Autogiro Company of America and Pitcairn joined the board of the British company at Hanworth. Pitcairn was a man of great honour and integrity, upholding Cierva's work and principles at all times. Additionally he strove to protect Cierva's patents, all of which Cierva had made over to the American before his death. Soon after it had gained its C of A, in August of 1929 Cierva packed up his latest machine, the C.19 Mk.II, G-AAKY, and sailed with it aboard the White Star Liner *Majestic* to America to demonstrate it to his friend. Reassembled at Pitcairn Field, Bryn Athyn, Pennsylvania, the exhibition also served as a photo-call for the US press corps. Here on 27th August 1929 Cierva is pictured demonstrating how the new purpose-built Autogiro can spin up its rotor for take-off by directing the propeller slipstream up into the rotor by moving the tailplane and elevator together to form a continuous upwards-pointing surface. Pitcairn, far left, looks on while the crowd of journalists and hangers-on in the background absorb a tall-podded plate camera and, far left, a newsreel movie camera atop what we would have called a shooting-brake.

The Pitcairn-Cierva Autogiro Company of America was formed in February 1929 as a subsidiary of Pitcairn Aviation, Inc, and from this came the 'PCA' designation. The first machine built was the PCA-1 which was based on the Cierva C.19 Mk.1 complete with deflector tail and tandem cockpits. The PCA-2, however, was a wholly-different machine. To start with the two-seater had just one side by side-seating cockpit so the load distribution was constant, and the tail assembly was conventional. The span of the fixed wing was reduced from 33 feet to 30 feet and the four-blade rotor was provided from England. Construction of the fuselage followed Pitcairn practice of welded steel tubes. With a gross weight of 2,750 lbs (raised to 3,000 lbs on production aircraft) the green and white prototype first flew in March 1930 with the markings X760W. The engine was the 225 hp Wright Whirlwind seven-cylinder radial. Demonstrated here landing right in front of tall trees on the Pitcairn land at Willow Grove, it is noticeable that the Cierva tail-skid design was preserved.

Four-bladed wire suspended rotors defined the Kinner powered PCA-2. The landing gear of these machines always looked clumsy and over-engineered. This was one aspect that the Kellett Brothers were able to address very early on in the design of their neat little K-2.

55

Pitcairn PCA-2, constructors' number B-25 carried the markings NC10791 and, as the property of the National Advisory Committee for Aeronautics, additionally NACA-44. This was still a four-bladed rotor wire-braced and with the blades joined by drag wires at the half-span position. The thin rod in front of the front pylon is the power take-off from the engine to assist in spinning-up the rotor prior to take-off. Always this clutch-driven mechanism had to be disengaged before take-off.

Another view of NC10791, the Pitcairn PCA-2 operated by the National Advisory Committee for Aeronautics. This shows that Harold Pitcairn's development of Cierva's design was appreciably larger than its British equivalent. It also had wings which helped offload the rotor in the cruise, the turned-up wing tips effectively increasing the overall dihedral of the wings. Notice also the rugged undercarriage. The rotor support cables are visible and they are attached to a short conical pylon extension which also served as an oil tank to lubricate the rotor hub bearings.

It is a characteristic of human nature that all things are cyclic – they keep coming round. When the Autogiro first appeared in the air, the sages proclaimed 'Well, you'll never loop one of those things!' Usually accompanied by that universal qualifier – 'It stands to reason'. Canadian pilot Godfrey W Dean, curiously forgotten today, was the first man ever to loop an Autogiro when he took CF-ARO, a Pitcairn PCA-2, over the top at Willow Grove on 13th October 1931. He repeated the feat on numerous occasions to the open-mouthed masses of 'standers to reason'. His achievement was followed by others, but not too many. Twenty years later, a fresh generation of sages saw the helicopter and said 'Well, &c, &c…'

Pitcairn PCA-2 NC10787 was owned by Walter Hoffman of Santa Barbara Airport in California. Here is his machine parked in a field at Nantucket, Massachusetts, – just about as far away from home as one could get. Unfortunately, this States-trotting giro was later destroyed by fire on 3rd May 1932. Several machines were lost through ignition caused by an engine backfire and the on-board extinguisher was not up to the challenge.

Hand-swinging a Pitcairn PCA-2 – or perhaps just blowing out: he is about to turn the propeller backwards. This machine was appreciably larger than its Cierva equivalent and was also much more powerful. A three-seater as compared to the two-seat Cierva machines of the time, the front cockpit was on the centre of gravity and was the location of the payload – two people seated side by side or an equivalent weight in freight or cargo.

A Pitcairn PCA.2 comes in to land on the apron at San Diego in the winter of 1931. The wide stance of the undercarriage is apparent from this view.

This Pitcairn PCA-2, NC10781, was operated by Giro Associates as a crop-duster where it was photographed tied down for the night by America's leading autogiro test-pilot and operator, John McDonald Miller in 1938. Miller had a remarkable career first with Pitcairn, then with Kellett and, in 1931, he set a record for the first trans-American flight in an autogiro, setting a record unbroken for 72 years. Miller died in 2008 aged 102.

Every year, America awards the innovator who has made significant progress in aviation with the presentation of the Collier Trophy. This award was instituted by US publisher, sporting flyer and first president of the Aero Club of America Robert J Collier who died in 1918. On 22nd April 1931, Pitcairn pilot Jim Ray landed the prototype PCA-2, NC760W, to the White House, Washington, where, by invitation, he landed on the lawn. In this picture we see Harold Pitcairn shaking the hand of US president Herbert Hoover (centre) on the occasion of the presentation of this prestigious award to him. Left of the picture, behind the trophy, stand Pitcairn's executive directors Agnew Larsen and Edwin Asplundh. The technology of the microphones in front of them is interesting.

It is 13th June 1931 and three unknown observers watch an unidentified Pitcairn PCA-2 Autogiro fly past.

Probably unique in the history of the Autogiro anywhere in the world was this American 'Autogiro Port' set up by Earl Spangenburg Eckel near Washington, New Jersey. Here Eckel (left) poses with Harold F Pitcairn (with tie) at the conclusion of a flight which started at Willow Grove – Pitcairn's own airfield – and ended at Eckel's estate. This evocative photograph of a moment in time was taken by Donald A Eckel. The sign on the façade says 'Size 100 x 300 yards: Erected 1931: Pitcairn Autogiros'.

Just outside Washington at Warren Field in September 1932 a few aircraft gathered in the sun-drenched grass at what would one day become Washington's main airport. The attractions that far-off day included this Pitcairn PCA-2, NC10780. It was owned by America's popular female flier of the age, Amelia Earhart, and later went on to have a busy career with Beech-Nut Packing and finally with Gyro Ads, Inc, an aerial advertising company.

Bearing the name of Santa Barbara Airports Ltd on its fixed fin, NC10787 was a Pitcairn PCA-2 powered by a 160 hp Kinner radial engine. Its pilot was one Walter Hoffman of Santa Barbara Airports but, on 3rd May 1932, his worthy Autogiro suffered a backfire and was destroyed by fire.

It is clear how friendly were the relationships between the Autogiro Company licensees that they all worked in a spirit of close co-operation – truly the ethos established by Cierva himself. It is 24th April, 1931, at Philadelphia Airport – the proud day when the Kellett brothers rolled out their prototype K-2, later to be NC10766. Here the diminutive machine is almost hidden behind the Kellett workforce. The tall chap in the centre and directly in front of the windscreen is Elliott Daland, formerly half of the Huff-Daland aircraft company. To the right of him is Chuck Miller. The front row, third from left, is test pilot Guy Miller, Wallace Kellett and Rod Kellett. Leaning on the wing is Harold Pitcairn and to his left are the Ludington brothers, Charles Townsend and Nicholas Saltus. Unfortunately, the Kellett's enthusiasm was in due course to cause a rift with Pitcairn.

The Kellett Brothers of Philadelphia built their first autogyro late in 1929 as an 'in-house' venture: it did not fly. In 1931 they formed The Kellett Autogiro Corp and obtained a Cierva licence from Pitcairn. Now with the experience of Cierva via Pitcairn behind them, they came up with the K-2, also known as the KA-1, first flown by Jim Ray on 24th April 1931. Designed in close co-operation with the Autogiro Company of America's design engineers under Wynn Lawrence LePage, the youthful British technician from the National Physical Laboratory, this first successful Kellett machine was powered by a 165 hp Continental A-70-2 or 210 hp Continental R-670, both radials. The K-3 was a cabin development of the K-2. NC10767, pictured here, was converted from the second K-2 to be built and fitted with a 210 hp Kinner C5 motor. The four-blade cable-braced 41-foot diameter rotor was mounted on a tripod pylon which would develop into the characteristic Kellett single-strut attachment of later years. Engine-powered pre-spin was fitted as was an engine self-starter using a Heywood compressor and air-bottle. Exhibited at the Detroit Air Show in 1932, this was the first of six that were built – four from scratch plus another conversion. This one, NC10767, was later owned and flown by the Steel Pier Company.

This is the prototype Kellett K-2, NC10766. Designed in close co-operation with The Autogiro Company of America, it was the product of a team lead by W Lawrence LePage, chief engineer for Kellett. The four-blade cable-braced rotor was 41 feet in diameter and of conventional Cierva design although of increased blade area. A mechanical rotor pre-spin drive was fitted. This, complete with clutch and gear box, weighed just 45 lbs. Additionally there was a rotor brake and a Heywood air compressor for engine starting. The first flight was undertaken by Jim Ray on 24th April 1931 and on 30th December that year Juan de la Cierva himself flew it at Philadelphia.

The diminutive Kellett K-2 was powered initially by a 165 hp Continental A-70-2 and later by the 210 hp Continental R-670 radial engine. Cleverly designed as a side by side two-seater and described as being 'more Cierva than Pitcairn', it was easier to fly than the tandem-seater Cierva or Pitcairn, had a simpler undercarriage than the Bryn Athyn product, and was fitted with a tailwheel as standard. Here NC11683 shows its attractive lines. It was operated by Ludington Flying Services, a business run by the Ludington Brothers, directors of Kellett, but in September 1935 it was sold to Argentina where it became R-287 existing until 1938.

A fine study of Kellett K-2 prototype NC10766 in flight. The Autogiro had a useful load of 609 lbs and a top speed of 100 mph with a cruising speed of 80 mph. Minimum flying speed was 24 mph. In all a dozen of these were built. One example, modified to K-3 status with a 210 hp Kinner engine and cabin top, went with the controversial American explorer Richard Evelyn Byrd (1888-1957) on his second expedition to the Antarctic (1933-35) and was thus the first rotary-winged aircraft to fly in Polar Regions. Originally marked NR12615, this was now NC12615 and gave yeoman service meeting its end under unusual circumstances. On 28th September 1934, having been parked outside in a blizzard, it took off in the usual sub-zero conditions, climbed to 75 feet then fell to the ground and was destroyed. Afterwards it was found that undetected drifting snow had packed into the rear fuselage and this had shifted the centre of gravity too far aft.

An artistic photograph of the prototype K-2 flying low over a golf course. A beautifully-composed image that's rather too good to be true. It is clearly a publicity shot to be used in advertising and, of course, it is faked! The golfers play on oblivious to the apparent presence of a flying machine a hundred feet above their heads. And notice the composition: golf flag central, Autogiro also central and superimposed on a perfect cloud. Never underestimate the abilities of the old 'wet-film' picture-makers to manipulate their shots long after the event!

This photograph, by contrast, is genuine and shows Jim Ray landing the Ludington's K-2 NC11683 on a golf course surrounded by men in plus-fours toting cloth bags.

At this time, America's sporting gentlemen played golf or enjoyed a chukker or two of polo. Thus promotional photographs for advertising purposes frequently associated these two. Here's Kellett's K-2 prototype NC11686 on the Menlo Circus Club Polo Field in California in the summer of 1931. One Captain Pierson poses with other members in front of the Autogiro as Samuel Metzger disembarks.

It is the spring of 1932 and Kellett K-2 NC11691 poses head on for the camera. Note the simplified undercarriage, the wider-tread wheels and also the larger blade area. The fuselage is wider than that of the comparable Pitcairn to accommodate side by side seating. The engine is a 165 hp Continental A-70. This aircraft was sold to Argentina in June 1935.

A historic snapshot of Kellett K-2 NC10767 taking off from a Washington street in 1931. Note the steep rate of ascent. This machine went on to complete an exhaustive series of US Army evaluation trials in 1931. By one of those quirks of fate, this aircraft was discovered derelict in a barn in 2002. It has since been lovingly restored to flying condition by its new owner, Al Letcher of Mojave.

An interesting view of the attractive little Kellett K-2 which had an initial rate of climb of 650 ft/min and a 250 mile range. While Pitcairn's PCA-2 carried a selling price of $15,000, this machine was priced at $7,885 as seen here or $8,255 with coupe top. This particular aircraft was sold to Argentina in July 1932 where it became R-237. It was withdrawn from the register before the end of 1938.

Down by the docks in Philadelphia, two Kellett K-2 Autogiros stand attracting mild attention from the public. In the background is a standard machine, NC11683, which was sold to Argentina as R-287 in September 1935 but in the foreground is the first K-3, a modified K-2 with bulbous cockpit canopy. This is NC10767 owned by Howard E Quick.

Prototype Kellett K-2 showing the spacious side by side cockpit and details of the rotor pylon – one main shaft leaning back braced by two rear struts. The forward shaft is the pre-spin drive from the engine. The wide-track undercarriage is seen to advantage here. This machine, NC10766, went on to become the property of The Steel Pier Company.

A gathering of the latest hardware: five Kellett K-2 Autogiros. Nearest the camera is NC10767 before it became a K-3. Third is NC10766, the prototype K-2 now painted up with the logo of its operators – the Steel Pier Company. The others are unidentified. The date is no later than the late summer of 1931.

Veteran giro pilot John Miller operated America's only mail service flying from the roof of Philadelphia's Thirtieth Street Post Office building to the airport at nearby Camden, New Jersey using Kellett KD-1B NC15069. The service ran from 6th July 1939 to 5th July 1940. The event is pictured here at the roof top landing site on 8th March 1940. In this group, Miller is fourth from left, the others being postal service managers Dalpiaz, Hall, John P 'Skip' Lukens and, to Miller's left, Leslie Cooper.

The same Kellett pictured on arrival at Camden on that March 8th with postal workers, left to right, 'Slim' Wenske, Holger Hoiriis and 'Whitey' Whitesell. Both these pictures came from 3-inch square black-and-white negatives in Miller's private collection which had survived damp storage. An interesting observation is that these two images were selected from the double images taken using a twin-lens stereo film camera, already something of an outdated format when these were exposed.

History in the making at the International Amphitheater at the Chicago Stock Yards on 27th January 1938. A Kellett YG-1A Autogiro owned by the US Army Air Corps prepares to land in a field next to the building where an International Air Show is to be held. The aircraft, which was being flown by veteran Autogiro pilot Lieut John Miller and Captain E C Hill, was to be part of the display which opened the following day.

The Kellett Autogiro Company was, like Harold Pitcairn, loyal followers of Cierva but if their first Autogiros were closer to the Hanworth models than even Pitcairn's, they actually progressed faster than Pitcairn and their KD-1 became the prototype of the first direct control Autogiro to enter production in the United States. Here is the prototype machine seen unmarked at roll-out in November, 1934. Powered by a 225 hp Jacobs L-4MA, it had a 40-foot diameter three blade cantilever rotor mounted atop a stylistic single streamlined pylon. Military requirements mitigated against side by side seating and these later-product Kelletts reverted to tandem cockpits. This aircraft, marked X14742 (later NC14742) made demonstration takeoffs and landings in a Washington street and flew the first mails from there to Hoover Airport.

The military KD-1A demanded that both occupants had individual but equal field of vision, be it in the observer/reconnaissance mode or light combat/offensive role. The YG-1A was, accordingly, a tandem two-seater powered by a 225 hp Jacobs nine-cylinder radial. Ordered during the latter part of 1935 it was used by the US Army Air Corps. Here the folding cantilever rotor blades are seen locked into their tailplane stowage slots. Unfortunately this particular model was prone to 'the dance of death' – ground resonance – and the attrition rate was high. Worse, though, was to come and an in-flight hub-forging failure let loose one rotor blade at a height of 3,500 feet. The out of balance force instantly ripped away the whole rotor pylon and the pilot and observer were lucky to escape by parachute. Even so, the observer's arm was broken when, on the way down, the flailing wreck of the autogiro struck him in a final gesture of defiance.

73

We have seen how the formation by Harold Pitcairn and Juan de la Cierva of The Autogiro Company of America served to license the manufacture of Autogiros to the Cierva patents. The first licensee was, of course, Pitcairn, and the second the Kellett Brothers who went on to produce highly successful military Autogiros and, later, helicopters. There was a third and this was the Buhl Company who had as their designer the talented Étienne Dormoy (1885-1959). Many writers, even the respected historians, cannot spell names correctly: poor Dormoy (remembered for his 'Dormoy Bathtub' light aircraft if 1924) is one frequent victim of such carelessness. The Buhl Company only ever made one prototype Autogiro and it differed dramatically from those of Pitcairn and Kellett in that it was a tandem two-seat pusher intended for use as an aerial photographic platform. It made use of a Pitcairn rotor and gearbox plus undercarriage. A curious but clever machine, it was hardly suitable for production yet nevertheless it played an important part in the development of the United States arm of Cierva's work.

Opposite page, top: The Buhl is prepared for its maiden flight in the hands of James W 'Jimmy' Johnson. With engine running there's time for a last briefing and a quick couple of photographs. The Detroit News chase-plane is the PCA-2 NR799W and is seen running up in the background. The Buhl rotor had a diameter of 40 feet and a maximum take-off weight of 2,000 lbs. Lawrence D Buhl and Alfred Victor Verville formed the Buhl-Verville Aircraft Co in Detroit but in February 1933 the company entered receivership – a victim of the recession. Happily, the Buhl Autogiro survives and it is today in Stanley Hiller Junior's Hiller Helicopter Museum in California where it has been restored although sadly the names of its designer and co-builder are carelessly misspelled.

Opposite page, bottom: While Buhl fell victim to the 1932 depression, and fourth Cierva licensee F W Steere never even produced a prototype, Kellett went from strength to strength. This culminated with the 300 hp Jacobs-powered YO-60 jump-start Autogiro of 1941. Designed by Richard Prewitt for the US Army Air Force, this tandem two-seater cabin machine was probably the most advanced of all the Cierva derivatives. Seven were built, this one being 42-13609 with the US serial 213609. Their high cost combined with ground-resonance susceptibility restricted their use. It was the last Autogiro model that Kellett would make.

75

Aircraft with rotating wings – literally – are, perhaps fortunately, few and far between. The best recalled was the Herrick Vertoplane but there was at least one other. This was C L Stauffer's Gyroplane. Quite unrelated to anything from The Autogiro Company of America or Cierva, this peculiar machine was the product of an inventor about whom nothing seems to have been recorded. In February of 1931 these two pictures were syndicated from a Chicago-based news bureau under the heading 'Revolutionary stride in aeronautics'. The picture story tells us all we know: 'Capable of flying 145 miles an hour and yet able to land gently at 12 feet a second, the Gyroplane is the conception of C L Stauffer, of Elkhart, Ind. When the upper wing is locked, as in the top photo, the machine has the appearance of a biplane and is capable of greater speeds than the Autogiro. In taking off and landing, the upper wing is allowed to swing free, as in the photo below, giving the Elkhart inventor's 'plane all of the advantages of the Autogiro'. What can be determined is that the single-seat aircraft looks workmanlike with its Kinner 100 hp radial engine. The three-strut rotor pylon was a clever insight to the way in which Cierva, Pitcairn and Kellett would later go, However, the enormous rotating wing, which appears to be of constant form airfoil section (as, indeed, it would need to be if is was to serve part-time as a real fixed wing), would have generated the most fearsome twisting loads in rotation from unequal life not to mention shifting centre of pressure. As a suicide machine it was well-built – except for that fearsome top wing/rotor. Its fate is unrecorded: it is unlikely to have flown, even as a fixed-wing aircraft.

Gerard Post Herrick of Ohio watched the development of the Autogiro with interest. He was a lawyer who wanted to be an engineer and, above all, to prove his theory that the advantages of fixed-wing flight might be combined with those of the Autogiro to create a fast machine for distance flight but with the STOL capabilities of a gyroplane. With the former chief designer of Kellett, he designed his first convertible machine and founded the Vertoplane Development Corporation of New York to finance it. Amazingly it flew as a fixed-wing aircraft, also as a combined fixed and rotating-wing machine, but the first attempt at that crucial in-flight transition ended in disaster. The pilot baled out of the uncontrollable aircraft but his parachute failed to open.

Herrick persevered with his second model, pictured here at Fairchild Field, Farmingdale, Long Island on 18th May 1936. Now referred to as a Vertiplane (sometimes, even, as Vertaplane), it underwent long and costly development until on 30th July 1937 it successfully completed an in-flight transition. There was some interest from the US Navy but despite constant attempts at seeking funding and contracts, the project died with him in 1955. His dream machine is preserved in the Smithsonian Institution.

The Cierva company had changed direction by the end of the 1939-45 war. It had now put the softly-spoken Autogiro with its almost silent swishing rotor well behind it and gone over to the development of the helicopter, the driven rotors of which would increasingly form the most annoying noise pollution known to mankind. The W.11 Air Horse was built at Eastleigh as a 24-seat passenger-carrier or freight helicopter powered by a 1,620 hp Rolls-Royce Merlin 24. Two aircraft were constructed, the first of which, G-ALCV, was first flown by H A Marsh as VGZ724 at Eastleigh on 7th December 1948. This picture reveals just how dependent the aircraft was on the 100% integrity of its rotor system: only one rotor failure would spell instant disaster; which, of course, is what would happen. Note the extended fin surfaces reinforcing the argument that almost every rotary-wing aircraft of this period possessed insufficient fin area.

The Cierva Company's move into helicopters saw the creation of what was then the world's largest example. The Rolls-Royce Merlin-powered Air Horse was impressive in flight, curious on the ground and a step too far in engineering. Its mid-air structural failure took the life of Alan Marsh and his crew.

The Fairey Jet Gyrodyne was described as the first British tip-jet-powered helicopter. Because it was tip-jet driven it did not need an anti-torque tail rotor. And because it had wings it could offload the rotor. Closely associated with the autogyro, it was almost a pure helicopter. Developed from the original Gyrodyne with a mechanically-driven rotor (and a curious offset propeller on the tip of one of its little stub wings), the Alvis Leonides-powered engine was used to drive compressors which supplied compressed air through the hollow rotor blades to small jet engines at the extremities where fuel was mixed and ignited. In cruising flight, these could be switched off to provide a free-turning autogyro rotor while the engine power was diverted to a propeller on each wing tip propelled the aircraft forwards. At the time of its first flight on 8th January 1954 it possessed the largest metal blade rotor.

Fairey Aviation's Jet Gyrodyne was registered G-AJJP on 1st March 1947 and, having demonstrated its practicality as a compound autogyro/fixed-wing aircraft (along the way creating a large area of people with hearing difficulties), went to the MoD as XD759. Eventually it was pensioned off as XJ389 seen here parked on a concrete slab outside what might well be an ablutions hut: relegated to serve not as gate-guardian, but lavatory-guardian perhaps?

Fairey's next move was to apply the Gyrodyne principle to an enormous compound machine, part autogiro, part helicopter and part fixed-wing aircraft. Intended for city centre to city centre transport (which, in the late 1950s, was believed to be the future of air transport) the outcome was the Rotodyne. There was no doubt that the Fairey Rotodyne was an impressive sight in flight as this early photograph of the unpainted aircraft shows. Note that the upper fins are shown angled and that there is no central or third fin. It was promoted as the world's first vertical take-off airliner.

A later air-to-air picture shows the upper fins arranged vertically and the addition of the fixed third fin. The undercarriage, fixed down for all the preliminary flight trials, is now retracted.

Pictured here with its folding upper fins extended and the blades under power, Rotodyne taxies out for the cameras.

By July of 1961 the Rotodyne was well into its development programme. However, with the merger between Fairey and Westland Helicopters, the writing was more than just on the wall. Despite frantic efforts to drum up interest, nobody wanted it. Harold Macmillan's government decided the project was well overspent, and Westland put its money behind the helicopter division of Bristol Aeroplane. Rotodyne and its 48-seat city-centre to city-centre capacity was axed. This picture reveals the added central fin to the parallel-sided cigar-box-shaped fuselage.

The Fairey Rotodyne has generated more controversy than almost any other flying machine with the possible exception of the TSR2 and the V-1000. Even today there are those who claim that its loss was a mortal blow to post-war aviation development. In fact although it may be long-dead, with unfailing regularity it crops up afresh with schemes to revive the concept. It was an impressive sound when it made its Farnborough debut but it was outstandingly noisy due to its four rotor-tip jet propulsion nozzles.

Fairey Aviation went to great lengths to demonstrate its utility. This picture was taken on a press camera-call to illustrate how the capacious rear-fuselage clamshell doors could accommodate all sorts of military things. Nobody was impressed enough to buy one, however. In this view, the folding vertical upper fin extensions are shown in the horizontal position.

The Fairey Rotodyne captured the imagination of many rival designers who suddenly envisaged a whole raft of worldwide opportunities for city-centre transportation by compound gyrocopter. On 12th March 1969 Lockheed California released this artist's impression of its contender. The accompanying press release read: 'Combining the helicopter's VTOL capabilities with top-class performance as a fixed-wing plane, it could be operational in five years as a commercial and military proposition'. Unlike the Rotodyne, Lockheed's scheme went one step better: the three-blade rigid rotor allowed a vertical take-off but after transition to level flight the rotor was stopped and folded back into a containing compartment along the top of the fuselage converting the whole thing into a normal fixed-wing aircraft. At its destination, the rotor compartment opened like a top-surface bomb-bay, released the rotor – and the thing landed like a helicopter. Nothing further was heard of it.

Opposite: After the war, many people were inspired to take up the rotating wing principles and some even thought to combine the one with the other as a sort of helicopter-gyroplane. In January 1958 a former Messerschmitt engineer named Bode Franke revealed his three-seater 'Volks-Copter' or Frankocopter about which little is known except that it had contra-rotating rotors and a 105 hp four-cylinder engine. Described as one of the first rotary-winged aircraft to have been built in Germany since the War, it had an eight-metre diameter coaxial rotor and a projected cruising speed of 160 km/hr with a fuel consumption of 18 to 22 litres/hour. The designer said it could be built for 65,000 DM (at that time about $15,475) which was about 170,000DM cheaper than machines then on the market. Announcing that trials were due to begin immediately, the engineer from Hoffnungstal near Cologne poses proudly with a creation that somehow looks more like a tank than a helicopter – and probably just as lethal. Herr Franke's creation was not heard of again and the reason for the heavy weight suspended from the back end would remain a mystery for all time.

87

In 1959 a French company designed and built a prototype compound autogyro designed by Georges Lepère to enter a contest to make an affordable and docile two-seat club-type aircraft. Registered F-PJCO, the Helicop-Air L-50 Girhel was powered by a 150 hp Lycoming engine and had conventional wings. The idea was that it would take off as an autogiro and then, as the cruising speed built up, offload the rotor by using the wings. It was a good idea – on paper at least. Here it is seen at the 1959 Paris Air Show at Le Bourget in the company of a Hiller 'chase' helicopter. Unfortunately, the Girhel never flew and hence never needed chasing.

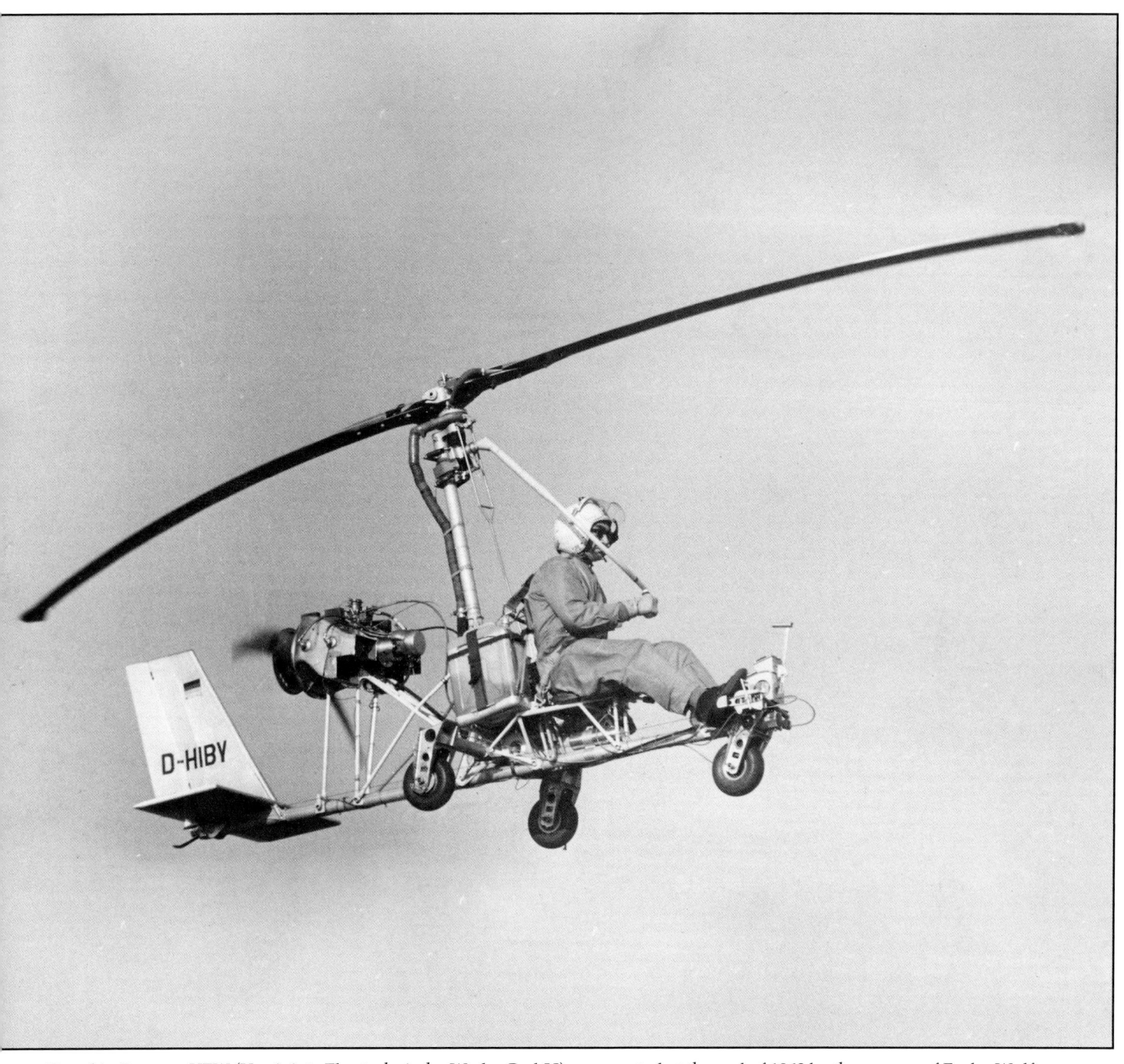

Based in Bremen, VFW (Vereinigte Flugtechnische Werke GmbH) was created at the end of 1963 by the merger of Focke-Wulf (which had developed a foldable towed autogiro glider for use on submarines as an extended surface lookout vehicle) and 'Weser' Flugzeugbau. The idea was, among other and larger things, to develop a family of light autogyros. The WFG-H2 (72 hp McCulloch four-cylinder two-stroke) was a single-seater designed along the Bensen lines but it took the Cierva science one step further in that this giro could actually hover. The fully-articulated two-blade rotor was equipped with tiny tip burners. While cyclic control was achieved in the normal fashion by tilting the rotor head using a direct control Bensen-type hanging stick, the usual collective pitch control was replaced by speed control of the rotor using the throttle. The engine drove a fixed-pitch Hoffman pusher propeller and also a small centrifugal compressor. For take-off, this air supply could be bled 'cold' to the rotor tips to shorten the spin-up time. However, by mixing this air supply with fuel from the main engine supply tank at the nozzle and igniting it, the rotor became jet-driven. A rotor speed of about 630 rpm was required for hovering while normal autorotation was 550 rpm. On ground tests, speeds up to 780 rpm had been attained. Bearing the civil registration D-HIBY, the WFG-H2 first flew in the autumn of 1966 and proved successful. At the end of 1967 the aircraft was grounded and work progressed to a larger three-seat version called the H3. This turbine-powered 'convertiplane' had a butterfly tail and a fully-enclosed cabin but development work did not continue. The registration D-HIBY has, confusingly, been re-used and has appeared on no fewer than four different machines. After the aircraft pictured here, the same letters appeared on a Bell 206B3, a Bell 206L and a MBB BO105D. Takes the art of recycling to new limits!

Raymond Umbaugh had been a fertiliser-maker but after experimenting with the Bensen Gyrocopter he moved into designing his own machine, Called the U-17, it was built for him by the Fairchild Engine and Airplane Corporation of Hagerstown, Maryland, and flight trials were carried out by Ken Hayden and one-time Pitcairn pilot 'Slim' Soule. Originally fitted with a 'T'-shaped tail, directional instability inspired the triple-tail solution resulting in the U-18 model. Here is another view of G-AYUE/N6150S taken at Membury Airfield. This former USAF base exists today as a private landing strip alongside the M4 Motorway some ten miles west of Newbury, Berkshire.

The Umbaugh U-18 autogyro takes to the air in its final configuration. Repeated troubles with lateral stability saw four different types of tail tried starting with a 'T'-shaped one, a 'V' and then reverting to a 'T' with large vertical end plates before the triple one as seen here was adopted. This had two fixed outer fins and a central all-flying rudder fitted to the fixed tailplane. A manufacturing agreement with Fairchild was combined with the formation of a dealer network, each of which was required to pre-order a sales quota. The principle was that if it worked with fertiliser, then it would work for anything, autogyros being no exception. In didn't, dealerships withdrew, the Fairchild contract became a costly fiasco and the business was wound up in 1962. After being revived as Air & Space Manufacturing, various problems terminated in a second liquidation at which point many items, including finished aircraft, 'went missing'. Subsequent revivals as Farrington Aircraft Corp and finally as Air & Space America, Inc, also met with failure.

Here is a picture of Umbaugh's curious triple-finned side by side two-seater autogyro. The Florida-designed machine dates from the mid-1950s and was the first autogiro to be certified by the FAA (which replaced the old CAA in America). The Umbaugh U-18 had a turbulent career and at least three names including the prosaically hopeful 'Air & Space America U18-A'. The problem was the designer's over-hype, under-financing and wishful thinking. Production began in 1959: by 1966 the firm was bankrupt. Three examples appeared on the British civil register; the first was G-ATZT but in the end the aircraft was not imported and the marks lapsed. Next was G-AYUE, pictured here still in its original US markings as N6150S. This was evaluated by Campbell Aircraft Ltd at its Membury Airfield where this picture was taken in September 1970. Unfortunately, on 21st May 1971, the aircraft was damaged beyond repair in a landing accident. The third example was G-BALB, formerly N6170S. This actually flew trials in and out of London's Battersea Heliport on the Thames but its main disadvantage was noise. It was withdrawn from use at Biggin Hill on 16th May 1974.

The Campbell Cougar two-seater was an attempt to make a side by side two-seater gyroplane to rival the Umbaugh machine. Only one example was built, G-BAPS of February 1973. The founder of the business was 'glider-doctor' Donald Campbell with the initial intention of handling the Bensen machines in Britain. This was in 1959. On 4th November 1967, he and his wife were passengers in an Iberia Caravelle, EC-BDD, which crashed at Fernhurst, Surrey, killing all on board. The business was maintained by his brother John but two years later it closed. Several subsequent attempts at revival have failed.

A British post-war attempt at a two-seat cabin autogyro was the Gadfly, the design of Eric Smith. Financed by Sqdn Ldr James Edward Doran-Webb of Thruxton Aviation & Engineering Company (the business which converted the elegant Tiger Moth two-seater into the obese Thruxton Jackaroo four-seater), the machine was also promoted as a potential crop-sprayer and known by several names starting with the Thruxton Gadfly, the ES.102 and, later, the Gadfly HDW-1. Registered G-AVKE, the engine was a 165 hp Rolls-Royce Continental IO-345-A. Exhibiting a quite crude standard of engineering, the Gadfly was shown in the static park at Farnborough in the mid-1960s and numerous modifications failed to inspire it to venture aloft. An age-old polite description for non-flyers of any type was to describe them as having air *in* the tyres, but none *under* them!

In post-war America, the autogyro underwent a renaissance as a means to cheap flying for the amateur and Igor Bensen designed a series of easy to build simple machines. His early models had a two-blade rotor directly controlled using a hanging stick. This model, with conventional joy-stick, was the B.7MC. G-APUV was built by Sqdn.Ldr R A Harvey, gaining its authorization to fly on 30th August 1960. Power came from a 72hp McCulloch four-cylinder horizontally-opposed target drone engine. It was withdrawn from use in October 1970.

As post-war autogyro building gained popularity in Britain, Campbell Aircraft Ltd of Membury in Berkshire built several Bensen models before Peter Lovegrove designed an improved version called the Campbell Cricket. Power was provided by a 75 hp Volkswagen flat four engine. Some 33 production models were produced between the end of 1969 and mid-1971. G-AXRC was registered in November 1969 and is seen here preparing for flight.

Helicopters have never been all that far away from Autogiros, at least not in the minds of the lay public and the myrmidons of Whitehall. After all, the invitation for Cierva to come to Britain in the first place came about because we had spent a whopping great sum on Vittorio Isacco's Helicogyre in 1928. This monster, built at Cowes by S E Saunders, was a transitional machine with driven rotors powered by no fewer than five engines – one each on the ends of the four rotor blades and one in the nose to pull the whole thing along. After extensive (and expensive) trials and no results, the machine was abandoned. Prior to that, RAE Farnborough had built and tested Louis Brennan's machine which had an engine above the pilot which drove airscrews at the blade tips. It flew inside its hangar but the controlled straight-line flight was no more than 200 yards. It crashed. At this point Britain forsook rotor tip drive until the Jet Gyrodyne of the post-war years. In America, however, Igor Bensen, however, re-kindled the notion in his B-4 Sky-Scooter two-seater announced in March 1954. A compressed air-driven model had allegedly proved the concept. Said to have been developed in conjunction with a US Navy HEPARS (High Efficiency Propulsion And Rotor System), this curious hybrid had a central engine and pullulating propellers on its rotors. Even with the aid of CORT (Crumpet On Rotor Tips), it was soon oblivionated.

British 'gyro-guru' Peter Lovegrove dispensed with Bensen's hanging stick in favour of more conventional controls. G-AVXB was built by Lovegrove at Didcot in June of 1968 as a B-8M powered by the larger 90 hp McCulloch 4318E flat four engine.

The 27th SBAC Show at Farnborough starred Campbell Cricket G-AXVK flown by Jeremy Metcalfe and seen here being demonstrated over the runway. This event was overshadowed by the accident to G-AXAR, one of Wing Commander Ken Wallis's WA117 autogyros which, before a large Friday crowd, crashed killing the well-known test pilot, 'Pee Wee' Judge then of Beagle Aircraft. During the day, Prince Charles was given the opportunity to sit in the Cricket and try the controls.

95

The post-war renaissance in the autogyro that started in America with Igor Bensen's rather basic Gyrocopter, has since reached new heights of style and perfection with numerous designers and makers around the world. In Britain, Wing Cmdr Ken Wallis was the inspiration behind a whole raft of amazing little rotorcraft starting in the late 1950s. When Beagle Aircraft was formed out of Auster Aircraft and F G Miles Aircraft, Wallis was invited to join the consortium. Wisely he demurred, stating that he did not mind Beagle promoting, building and marketing his design, but it would remain his property. The decade of the Beagle business ended in tears but Wallis escaped the company collapse and went on to greater things. His sleek and streamlined WA.117, seen here in a classic Ken Wallis 1961 demonstration around his private house, reveals just how far the Cierva technology had been evolved.